ONE TEAM
ON ALL
LEVELS

Stories from Toyota Team Members

SECOND EDITION

ONE TEAM ON ALL LEVELS

Stories from Toyota Team Members

SECOND EDITION

TIM TURNER & COLLEAGUES

CRC Press
Taylor & Francis Group
Boca Raton London New York

CRC Press is an imprint of the
Taylor & Francis Group, an **informa** business

A PRODUCTIVITY PRESS BOOK

CRC Press
Taylor & Francis Group
6000 Broken Sound Parkway NW, Suite 300
Boca Raton, FL 33487-2742

First issued in paperback 2019

ISBN 13: 978-1-138-43480-6 (hbk)
ISBN 13: 978-1-439-86067-0 (pbk)

Visit the Taylor & Francis Web site at
http://www.taylorandfrancis.com

and the CRC Press Web site at
http://www.crcpress.com

Dedicated to the memory of Eugene C. Keith

I will forever be grateful for you taking the time to find those students who would have been overlooked.

Dedicated to the memory of Bob Ditty

Team leader at TMMK and coauthor of this book.

Contents

The Toyota Precepts

1. Be contributive to the development and welfare of the country by working together, regardless of position, in faithfully fulfilling your duties.
2. Be ahead of the times through endless creativity, inquisitiveness, and pursuit of improvement.
3. Be practical and avoid frivolity.
4. Be kind and generous; strive to create a warm, homelike atmosphere.
5. Be reverent, and show gratitude for things great and small in thought and deed.

Foreword

In the darkest days of the recession, I had the opportunity to visit three Toyota manufacturing and assembly plants in Indiana, Texas, and Kentucky. It was February of 2009, and too many people can remember vividly the depths of the malaise at that time. The economy collapsed in October of 2008, when the financial system of the United States imploded from speculative money lending and profit mongering that triggered a global crisis. By February the economy continued in a deep funk, and it was not clear to anyone when we would start seeing some sign that we had at least hit bottom, let alone started to recover. I write this in September of 2010, and it is still not clear whether this will be a double-dip recession; unemployment in the United States remains over 9 percent.

At the small number of conferences and training programs where they were still using me back in the winter of 2009, I would hear over and over, "We had to make major layoffs." It was as if it was ordained by the gods that a deep dip in sales meant the need for plant closures and layoffs. Now for the shocker: when I visited the Toyota plants, they were humming with activity despite the fact that 30 to 40 percent of the people were not really needed for the level of sales at the time.

Those are two surprising facts. The first was that Toyota continued to employ thousands of people that they really did not need despite a 30 to 40 percent reduction in sales and quarterly financial losses. The second was that Toyota was keeping all these extra people very busy. People who were not working in production were actively engaged doing something or were practically running to get to someplace they had to be.

What I was seeing was the "Toyota Way" at its best. The reason that so many nonessential personnel were still coming to work and getting paid was that Toyota thinks long term about their investments in people. "Respect for people" is a pillar of the Toyota Way, and that means truly valuing the high level of skill acquired and the culture created over years of development. The second reason is the second pillar, "continuous improvement," which means never stop improving even in a down economy. In fact, the best opportunity for improvement is when business is slow and there are a lot of extra people available to do kaizen.

What I heard from managers in each of the plants was actual gratitude that business had slowed down enough that there was the time for the deep training and kaizen that frankly had languished during the boom times for Toyota. Once they got down to the business of training, they realized they had gotten weak at it because people were working ten- to twelve-hour days just to keep up with demand. The reason why they could afford these remarkable investments in the worst of times is because they had done what your grandparents probably advised—save for a rainy day when business was booming.

Okay, so what does this have to do with this book? Around that time I was contacted by Tim Turner, who was organizing team members at the Georgetown, Kentucky, plant to write a book recounting how joining Toyota had so positively changed their lives. The economy was so bad, and there were some Toyota team members who were feeling insecure about their futures despite Toyota proving they were committed to job security. Tim and his team members wanted to write a book to share with others what a remarkable company Toyota was and to hopefully improve morale in the plant. He was overwhelmed by all of the people who took the time to write down their stories and wanted to share them in their unedited words.

Who might publish such a book? Frankly I was not sure, but when I read the stories I was touched to the core. All of these stories were in different voices, but they all had in common tremendous gratitude for the deep respect that Toyota had shown them and the personal transformation they had experienced working for Toyota. This was why I dedicated my professional career to writing five books about the Toyota Way (and three more are coming). It was the true dedication to developing and respecting people, the heart and soul of the company.

Tim Turner and his colleagues decided to self-publish, and I was able to help a little by recommending to Productivity Press that they republish the book to get it more exposure. And there you have it—the real-life stories of people from many different walks of life who have all been touched by the Toyota Way. Perhaps it will bring a smile to your face. Perhaps it will make some rethink their commitment to developing people. In the best of worlds, someone out there who otherwise would have pulled the trigger and issued layoff notices will hang on to their people and invest in them instead. I can tell you that reading this book brought a lot of smiles to my face.

Jeffrey K. Liker

*Professor, University of Michigan
and author of* The Toyota Way

Chapter 1

Our History

Kosai, Shizuoka, Japan, 1867

This was the year that Sakichi Toyoda was born. He was an inventor who is actually referred to as the "king of Japanese inventors." The Toyoda family is a very humble family probably due in large part to the family story. Sakichi's father was a carpenter in Kosai, Shizouka, Japan. The creativity and ingenuity that he used while in the carpentry shop were followed and studied by his son. Sakichi's most famous invention was the automatic power loom that would stop automatically when a problem occurred. This was known as jidoka (autonomous automation). The idea of jidoka is the combination of machines and people to ensure great quality, and it is built into every Toyota product.

Sakichi had two children: a son named Kichiro Toyoda and a daughter named Aiko Toyoda. With his father's approval, Kichiro would eventually found Toyota Motor Corporation and would set the groundwork of the Toyota production system in the form of a four-inch-thick manual. He first began implementing these ideas in the loom works factory. He would resign from president of the auto division in 1948 and ask his cousin and close friend, Eiji Toyoda, to take over. Eiji had already proven himself to be a great leader in 1938, when he was charged with the task of overseeing the construction of the Toyota Motors facility near the town of Kiromo, which would later be called "Toyota City." This factory would later be known as the mother plant to all other facilities throughout the world.

He would lead Toyota through hard times and successful times until it would eventually be known as a worldwide leader in automobiles. He would lead us through the creation of Lexus, which would be known as a worldwide luxury brand, the first from a major Japanese company.

Toyota Motor Corporation would twice attempt to get into the American auto market. The first time was in the 1950s, when it introduced the Toyota Toyopet. This car would have small success in California, but it could not get sold throughout the rest of the country.

Throughout the 1950s, members of Toyota would visit the United States to see how the Americans could mass-produce products. Eiji would visit a Ford plant at the invitation of Henry Ford II. He was amazed at how the Americans could build 8,000 cars a day when the Japanese (Toyota) could only produce 2,500 cars in their entire thirteen-year history. He went back to Japan with the challenge to catch up with the Americans' productivity in three years. Toyota had to increase production and do it in a way to save space and money.

This challenge created a lot of excitement within the company. Many team members and managers would take on this challenge with great enthusiasm, and none greater than an employee named Taichii Ohno. Mr. Ohno was born in Dalian, China. He is now considered to be the father of the Toyota production system and lean manufacturing. He was a faithful Toyota employee, getting his start on the factory floor at the loom works and then moving to the motor company in 1943. He is a true success story and an example that any employee at Toyota can grow within the company to become an executive and leader. This is a huge key to the success of every Toyota facility. Each employee is empowered to create "kaizen" (continuous improvement) and then "yokaten" (share those ideas to all other Toyota facilities). We should never rest on our laurels and always look to improve.

Mr. Ohno liked to share stories in his teachings. The following story is an excerpt taken from his book, *Toyota Production System: Beyond Large-Scale Production*.

> Years ago, I used to tell production workers one of my favorite stories about a boat rowed by eight men, four on the left side and four on the right side. If they do not row correctly, the boat will zigzag erratically. One rower might feel he is stronger than the next and row twice as hard. But this effort upsets the boat's progress and moves it off course. The best way to propel the boat faster is for everyone to distribute force equally, rowing evenly and at the same depth. Teamwork is essential to our success.

In 1959, Toyota would build its first plant outside of Japan. The plant would be built in Brazil. At this point, Toyota would commit to someday build and design cars where they are sold.

As you read this book, you will see examples that the systems that he and others created, combined with the values and humble nature of the Toyoda family and the corporation itself, would create a company that would provide stable employment for nearly 68,000 direct employees across the globe.

These principles and values would help shape a new idea in the world that people from across the globe can work together for a common goal. In our company's case, it is to build a work culture that provides safe cars to transport people from every region of the world to their destination.

Georgetown, Kentucky, 1985

It was known that Toyota wanted to expand to North America. One company principle that Toyota believes in is that they build the cars where they sell them. In January 1985, Toyota Motor Manufacturing, USA, would be established, and the search began to find a site for its first American auto assembly plant. It was narrowed to five states in November, and on December 11, 1985, it would be announced that Georgetown, Kentucky, would be the site of the new facility that would employ over 3,000 Kentuckians.

Ground was broken on May 5 for the building of the factory. Fifteen suppliers would also begin showing interest in locating to Kentucky. In 1988, the state announced that over 120,000 applications from every county had been turned in to the state for a job at Toyota. Most Kentuckians did not have a lot of choices as far as employment went. I suspect this was one reason why Toyota decided to locate in our state. They were able to pick the very best workers from a state in need of a strong employer. Many people would relocate their families from all over the state. A lot of people were able to leave a region of the state that provided only the opportunity to work in the coal mines for a good-paying job.

In May 1988, the first-generation Camry would roll off our line. This car can now be seen in the Toyota visitor center located at our plant. In August, the company would break ground on a new power train plant. The company now directly employs over 7,000 workers at the Georgetown facility and later established Toyota Motor Engineering & Manufacturing North America (hereafter, TEMA) in Erlanger, Kentucky. This would be the headquarters for all North American facilities.

Toyota had many concerns with locating in North America. It had to find a state with a culture very similar to the company's beliefs and values. Georgetown had to be a success story.

The corporation and the state could accept nothing less than a great partnership. Mr. Fujio Cho (former president of Toyota Motor Manufacturing, Kentucky; hereafter, TMMK) said it best at our twentieth-anniversary celebration: "Twenty years ago Toyota took a giant step in North America. In the beginning we were anxious. We were not sure if we would be accepted in the U.S., and especially in Kentucky. But as it turned out, the people of the community and the state welcomed us wholeheartedly and with open arms."

Toyota now has seven assembly plants in North America. The Georgetown plant is considered to be the mother plant of a few of the North American facilities. This is something that we are proud of and take very seriously. It is our job to provide other states and our fellow team members across the United States with the support and leadership that were provided to us twenty-three years ago. The key to Toyota's success is that we see everyone as our customer. This starts with management of our company seeing the team members as their customers and the team members on the factory floor having a customer-first attitude. We realize that if our cars are not built with quality in mind, then we go backward. We are always looking ahead.

Georgetown, Kentucky, January 2009: The Recession and Downturn of the Auto Industry

In January 2009, the world economy was the worst it had been since the Great Depression, and uncertainty surrounded everyone, including Georgetown for the first time since 1988. Our plant was idle. A rare, uneasy feeling was surrounding the team members. Our president, Steve St. Angelo, would walk around the plant raising morale. He referred to it this way. He felt as if he was the captain of a ship: he said he was steering the ship, trying to make sure everyone was having a good time but keeping the focus on avoiding that iceberg that he knows is out there. This positive attitude from our executives along with the understanding of the situation among our team members would keep us afloat. The year 2009 would be a challenge due to the economy, but nothing could prepare us for 2010.

Washington, D.C., February 2010

Two congressional hearings would take place in February 2010 to look at Toyota quality and claims of unintended acceleration. A Toyota customer testified that her Lexus sped up in Tennessee and she could not get it to stop. It was a heartfelt testimony of a call to her husband. After the incident, the Lexus was sold to another family who did not experience any issues. Our government used our tax dollars to buy that Lexus from that family and to tear it apart. As of November 19, 2010, they have found no issues to prove that it had unintended acceleration. Toyota Motor Sales President Jim Lentz was then questioned. It was a rough testimony full of politically charged allegations and claims. Keep in mind that I am not a spokesman for Toyota. I'm just an American who feels like our representatives in Congress mistreated my employer and that they are using my tax dollars to hurt my family by attempting to destroy my company's good name and reputation. Enough is enough. Congressional priorities need to stay focused on our economy and quit looking for ways to take the attention off of the real problems in our country.

The second hearing would have Toyota Motor Corporation President Akio Toyoda and North American President Yoshi Inaba sharing testimony. I think a lot of Americans don't realize it, but Akio Toyoda could have said NO to Congress. He did not have to testify. He chose to because our customers deserved to hear from him. It was a proud moment to see him "step up to the plate." Congress was brutal. Some of them seemed to be having fun at the expense of someone in a tough situation, and they took full advantage of their power. When I watched the hearings, I couldn't help but feel sorry for him. He had a language barrier and a cultural barrier to deal with along with the Congress members' sharp questions and comments. It was a proud moment for me to see him in our nation's capital. You see, he wasn't just defending his company or his family's name. He was defending the honor of the 37,000 Americans who work for his company. Under the circumstances, he did a great job.

The following are stories shared by Toyota team members from all levels of the company. Some discuss the importance of the fourteen Toyota Way principles, how they are an outline to create a culture of servant-style leaders, and how those principles with the idea of lean manufacturing are combined to create a culture of constant improvement.

The rest are stories from team members who wanted to share how the company provides them with benefits and demonstrates the values that the Toyoda family has created within the corporation to provide us with a stable secure future. It is these values that empower the team members to work hard for Toyota Motor Manufacturing and create a sense of loyalty to the company.

Chapter 2

Our Beginning: Team Members Who Began the Success Story

There is a story of two Japanese men who were driving through Georgetown, Kentucky, in 1985 and stopped to ask for directions to Cherry Blossom Way (the address of our plant). They were treated rudely and sent on their way. It would have been easy for Toyota to have gone back to Japan and said no to Kentucky. Thankfully, they did not do that. They realized that putting a plant in the United States was about breaking down racial boundaries. Diversity actually began in our plant and community before ground was even broken to build it. Georgetown would never be the same.

"It was an exciting time, working at the paper, waiting to see if Georgetown, Kentucky, would be the chosen one."

Debbie Poe: Assistant Manager

"Oh, What a Feeling!" is what the headlines read after the finishing touches were made late that night at the newspaper back in 1985. It was an exciting time, working at the paper, waiting to see if Georgetown, Kentucky, would be the chosen one. What a feeling it really was, to get the wonderful news that TOYOTA was coming to Georgetown and be able to pass it along to our readers.

From then on, it seemed that every edition had new and exciting news about Toyota. But not all of it was positive. We were really surprised that there were so many in the community that did not want the plant to be built here. What would happen to our community? They wanted to keep our small, hometown atmosphere. But there were so many more that wanted the growth that Toyota would bring. Growth had already begun, with the years in planning for a new, larger airport, which had to look for another site to build after Toyota decided on their new site.

Yes, it was a very exciting time for me, also. My husband, a pilot and the owner and operator of Fixed Based Operation, Marshall Field Airport, had hosted several meetings at the old airport during the planning phase of the plant. All of this time, he kept telling me, "You really need to apply for a position at the Toyota plant when the time comes." I would just say, "OK, I will think about it," all the time thinking to myself, "I have no skills to build a car!" I could barely identify the major components, much less build one.

The day came when there it was, in black and white ... I was actually pasting it to a page in the newspaper. "Toyota: now accepting applications." Everyone was talking about it. But what would I do there? Again, I don't know anything about building a car. Finally, my husband talked me into it.

I will never forget the feeling I had when I returned home one afternoon and had a message on my answering machine. It was from a member in HR at Toyota. The message said exactly this: "This message is for Deborah. This is Todd from Toyota. Please call me Monday morning at this number. I have some good news." YES ... Monday morning. This was Friday afternoon, after everyone had gone home for the day. I had to wait the entire weekend before I could speak to anyone. I was so excited. I remember screaming when I heard the message. My sons ran into the kitchen to see what I was acting so crazy about. I was so happy, I cried and screamed and explained to them that I thought I had been accepted at Toyota. The first thing they said was "But Mom, you don't know anything about building cars, do you?" I laughed so hard, then, we all started to scream. It was one of the happiest times of our family.

I started out my career in conveyance. My first job was picking up kanbans (a sign board for reordering parts) on a tricycle, every hour, then sorting them for the delivery routes.

I learned about the just-in-time delivery system of the Toyota production system (TPS) during my first few months on the job. I did the same job, every day, for approximately six months, until we began job rotations, in which everyone in my team got to do every job.

I then decided to broaden my horizons. I took team leader classes to prepare myself for the next level, and quality circle classes to become a circle leader, teaching and guiding my team on problem-solving issues in our area.

My next step was group leader, another highlight of my career. My manager at the time, Will Allen, took me to the shipping line, making me think that I had done something wrong, but with his kind words and gracious comments, he told me that he was honored to offer me the new position, if I would take it.

Today, my position as assistant manager in the assembly plant offers challenges every day. During my career, I have had the opportunity to work in every area, learning, teaching, and practicing TPS, including the plant in Canada and the Subaru plant in Lafayette, Indiana, and I have traveled to Japan several times for additional training.

I have also had great opportunities to be involved in the Diversity Advisory Council, established in 1993; the Assembly Diversity Committee; and the first business-partnering group at TMMK, Women's Leadership Exchange Network (WLEN), of which I am currently the chairperson. We chose to form this organization to support TMMK's success by encouraging, developing, and supporting through information and education an environment that recognizes and respects diversity in the workplace. Our objectives are to advance the recognition of women through acknowledgment of accomplishments and achievements; cultivate respect of women through education, leadership, and skill development; expand the value of women through providing opportunities to develop productive and long-term relationships;

and provide opportunities for networking and educational development. Our president, Steve St. Angelo, has been very supportive in all of our initiatives, and our vice president, Cheryl Jones, is the champion advisor for our business-partnering group.

I consider myself to be very fortunate to work for a company that believes in the development of their team members. Toyota is also committed to ensuring each and every team member has the opportunity to work in a positive environment, where everyone treats the other with dignity and respect. At TMMK, we strive to be an inclusive organization, where each team member is recognized, understood, and valued for their unique personal and cultural differences.

I could not have imagined that I would ever work for such a great company as Toyota. My entire family has benefited more than I could have ever imagined from my career over the last twenty years. My dear father passed away when I was nine years old; I grew up not having much. As a girl originally from a small town of population 300, my Aunt Geneva has told me that my dad would be so proud of me. To have been able to have the experiences that Toyota has given me has been priceless, not to mention the training that I have had, actually building a car, after all! That makes all the hard work worth it.

"Oh, What A Feeling."

> Debbie Poe is currently the assistant manager in our North American training center. She is a highly respected assistant manager and a really great person. Her hard work, dedication, and teamwork in groups such as WLEN and the Diversity Committee have brought forth great opportunities and understanding to many team members. —Tim

"1988—a large factory is locating here in Scott Co.... This area will change forever!!! That was the thought of locals here."

Karen Wells: Assistant Manager

I am a native of Scott County. My ancestors were some of the first settlers in Scott County, settling in a place called Turkeyfoot, named by surveyors because the three creeks come together to form the shape of a turkey's foot. It's about six miles from the plant.

There were some uneasy feelings when talk of the Japanese building a plant here started. Everyone watched as a huge building began to take shape. Then the hiring process began. It did not take long for the locals to embrace the newcomers as friends. They soon touched the lives of most families in the area. They worked jobs generated by outside suppliers, groundskeepers, cafeteria workers, and the list goes on. These workers became part of TMMK's extended family, improving the quality of life for many.

The teachings of Toyota quickly became part of our ways.

A few that come to mind: "If the student hasn't learned, the teacher hasn't taught." OR "Constant kaizen—always improving keeps us sharp, and sharing our knowledge with others helps them improve and raises the bar for us." AND "The most important part of a business is the workers [TMs], not management, and they are treated as such."

So, the thoughts twenty years ago that TMMK would change us forever were very true—all in a positive way. The financial benefits are huge, but so are the lessons learned that are now a part of many of our Kentucky daily lives.

Karen Wells retired while this book was being written. I had the pleasure of working for Karen for a couple years. I told Karen one time that I figured out her leadership style. She was in charge of the chassis section, which was predominantly men. We all saw her as if she was our favorite aunt. No one wanted to let her down. She was a tough manager when she needed to be, but she didn't micromanage. I have been told that this place is so big that no one will be missed. I will have to disagree with that statement. Karen Wells, you are missed, and I hope you are enjoying retirement and planting your flowers. You deserve all the happiness in the world. I learned a lot from you. —Tim

"I have never been more proud to be a Toyota team member than I am right now."

Keith Royse: Group Leader

The commute to my job in Lexington became more than routine as I drove by the construction site of the new Toyota plant outside of Georgetown, Kentucky. As the roads were being completed, and the activity around building a facility of this size was going on, many people that I knew were applying for the jobs at the new Toyota plant. This is the site that would soon be the home of the Toyota Camry. In those early years, the people that I knew had a lot of pride in the fact that they were working for Toyota.

At the time the Toyota plant was starting production, I had a good job and had no reason to be looking for another one. However, as time went on I kept hearing everyone discussing their jobs at Toyota and that there were a lot of opportunities there. I started to think about applying for a job at Toyota. In 1988, I thought, "Why not—I'll apply and see what happens." From what I was hearing, the chances were slim that I would get a call anyway. The hiring process was selective, and numerous people were applying for the jobs. After filling out the application, it wasn't long before I was called to come to the employment office in Georgetown for some tests. A few weeks after these tests were completed, I received a call that informed me that I was to attend some assessments in Frankfort, Kentucky. These assessments lasted two days, and when they were completed I thought, "It sure was a lot of work just going through the hiring process for a job at Toyota." I felt like I had done well, but I wasn't sure and the doubt set in. It was some time before the next call came, but it did and I was told that I was being placed in a hiring pool. I would be contacted for the next step in the hiring process. Well, time passed, I was busy with my job and family, and didn't think much about a new job or leaving the one that I had.

Now fast forward to the year 1990. In 1990 and early 1991, there were several changes in my life. The company I was working for was going through a major change and was bringing on a new product line. I was one of the production leaders assigned to the change-over facility where the new line would be started. The change-over team spent a great deal of time traveling to get ideas for the new facility. One of those trips was to the headquarters of the company that I worked for. This trip lasted for thirty days, and during this time the team was taught the basics of the Toyota production system. I have been at Toyota for seventeen years now, and I am still learning. As I returned from that trip, there was a major change coming in my life that I wasn't prepared for. My father was ill, and when I returned he was placed in the hospital, a trip that he wouldn't return from. After a time of reflection and looking at where we were and where we wanted to go, my wife and I decided that I should pursue a job at Toyota one more time.

It was late 1991, around mid-October, that I got the number for Toyota from a friend. While on the phone with the representative, I was told that there was no file on me and that they had no information on my hiring status thus far. After going over the status letter that I had received years earlier, the representative told me she would look into it and gave me her phone number to call the next day. After I hung up the phone, I thought that I had waited too long and my file had been discarded. I felt that I would have to start the hiring process again or just drop the idea altogether. I called back the representative the next day at the agreed-upon time, and she told me that my application had been misfiled and that it had taken her quite a while to find it. Now that it had been located, she had set a time for me to come to the plant for an interview. Wow, was that fast! I was impressed that she had spent so much time trying to help me when she could have just dropped it. After the interview, a physical was scheduled, and the results were ready after Christmas. Approximately three months after talking to the Toyota representative, I received the call that I had been waiting for. I was to start work at Toyota on January 28, 1992.

No one knows what to expect when starting a new job, but from day one I was put at ease and given all the support that I needed to succeed. My group leader, David Everly, showed me around, introduced me to the group, and answered my questions. My first few days on the job involved a lot of training. Each process was covered step by step until I had them mastered. Everyone offered some sort of support, which made the transition easy. There was a lot of pride in the work that was being done, and people were proud to be working for Toyota. I was taught that I could be involved in changing anything that would improve the processes. That was a new concept to me, and I decided to join a quality circle. A quality circle is when a group of team members get together to work on a problem.

I was proud to be working for Toyota. I spent the next two years on second shift in chassis line 1. During this time, I enrolled in team leader classes to learn more about the Toyota production system and what was expected at the next level. Each new job assignment brought its own set of challenges as well as its rewards.

After seventeen years at Toyota, I have seen many changes and have been involved in some of them myself, but the one thing that has not changed is Toyota's commitment to its team members. Over the years, I have heard over and over again that Toyota is in it for the long term, and with the current economic conditions I am seeing that commitment played out right before my eyes. With all the turmoil in the world economy, Toyota is doing everything it can to keep the company strong and to maintain long-term employment security for its team members. I have never been more proud to be a Toyota team member than I am right now.

> Keith has worked in many areas of the plant. He does a great job, he cares about his team members and their well-being, and for that I thank him. —Tim

"To this day, as they did for me, a ribbon is tied to the front entrance tree to the plant that signifies a team member is deployed."

Tony Hendrichs: Group Leader

My Story as a Military Team Member at Toyota

I started at the Toyota Georgetown plant in Kentucky in 1989. As most in the shop, I had no automotive production experience. I knew the job and the benefits would be my best opportunity at that time of my life.

I signed up for the Kentucky Army National Guard fresh out of high school. This was my only option to help pay for the college I wanted to go to, Western Kentucky University.

I was very apprehensive about being employed at Toyota and having to stay enlisted in the National Guard. I heard some stories of other soldiers in my unit, about the lack of support or job losses due to the required two-day and some three-day weekends that we had to attend. I quickly informed my group leader about my requirements on my drill weekends. I was hired for second-shift work, so on the drill weekends, I would have to report to my unit in Danville, Kentucky, by 7:00 or 8:00 a.m. after getting off of work around 3:00 a.m. prior that night.

After being at Toyota only a year and a half, I was ordered to report to active duty in October of 1990 to support Operation Desert Shield, which later grew into Operation Desert Storm. At that time, I was the first team member to be deployed while working at the plant, so many of the specifics of the military support policies were not decided or challenged. I was told before I left that my job was secure and would be held for me until I returned. I left, leaving a wife and a two-year-old boy behind, wondering what my future in the military and at Toyota would be.

Once deployed, I received letters at home from our Toyota plant ensuring my current job position and pay would remain until I returned. While there, I received many letters of support from local schools, businesses, and my fellow team members in my work group. Along with many care packages, this kept me encouraged to perform my mission with my unit.

I also received the monthly company newsletters and updates on what was going on with the plant, which would also help me keep up with what was going on at home and with work and others.

Eight months went by, and I had finally returned. While I was deployed in Saudi Arabia, I sent many the favorite local dollar bills the Saudis used, plus obtained some unique Iraq money, as souvenirs. I sent a riyal, a Saudi dollar bill, to Mr. Fujio Cho, our TMMK president then, as a small appreciation, with a letter thanking him for supporting my family through my absence from work. I also received a surprise check one month later for my wage increase while I was gone, like I had been there every day at work.

I found out, after talking to and hearing many fellow soldiers, that I had the best of most employers whom would support my family and me during my deployment. Many soldiers came back to unemployment and lost wages while they were gone.

Since then, I have finished my enlistment with the Army National Guard, but still support and stay involved with the military support programs we have at the plant. I am also used as a resource for new policies and guidelines regarding military leaves and deployment.

To this day, as they did for me, a ribbon is tied to the front entrance tree to the plant that signifies a team member is deployed. Once the working soldier returns, a small appreciation ceremony removing the ribbon is held.

If any company could learn how to support their military employees, I believe TMMK is a great example.

> Tony has been my group leader on two different occasions: once in production and now in my current assignment in the assembly safety team. Both have been learning experiences for me, and I'm honored to call him my friend. —Tim

During the writing of this book, I received an e-mail. A team member had recently retired and had written a letter to our executives. I immediately went into the mode of trying to find this person. This letter needed to be shared with our customers.

To whom it may concern,

There are many moments in life that we carry around in our memory bank to remind us of a life well lived. Being the mother of six children and grandmother to five I certainly have my share of those memories. One of the highlights of my life was receiving that letter offering me employment in a company with new ideas and a chance to be part of a team. Even though we had one plant and one shift back then (1988), I just knew I would not be able to find my group on the 2nd day of work because it was the largest building I'd ever been in. Learning about "just in time," "kaizen," putting out "fires," changing forklift batteries and of course how to put down yellow tape everywhere became "habit" after a while. Meeting our trainers from Japan and learning about other cultures was just a bonus. Next came my greatest source of PRIDE: getting my team leader hat. The first time I put it on, I felt 10 feet tall. I am not a hat person, but that hat meant something. The opportunity to solve problems with my team and feel that I made a small difference meant a lot to me. As I have aged and have had to learn how to cope with failing health, it is what I learned from working for this company that has helped me the most. Keeping things in their place or (5's) helps me find my way in life as my eyesight continues to fail.

Thank you Toyota for the opportunity to help my family, myself, and, in some small way, the success of Toyota.

Sincerely,
Cheryl Jackson Clark
TM #1575

"The offer came on Good Friday. I really do not have
the words to express what that meant to me then
and what it means to me now."

Audie Alford: Group Leader

The first time someone suggested that I apply for a job at
Toyota, I replied, "Why? I have a good job." That was just one
of the times my wife proved she was much smarter than I was.
This was in early 1988, and I was going through what I thought
was a secure stretch of employment. I had worked twenty
straight months without being laid off. Looking back, that does
not look so secure. I was laid off again shortly after that.

I went to the employment office in Corbin, Kentucky, and
applied for a position at Toyota. I filled out the initial paper-
work and then went home and waited. Some time later, I was
called in to take the first test. Later I was called and scheduled
a series of testing at the Frankfort Civic Center. This was called
Project T. I was very excited and also very nervous. I grew
up in an old mining camp in Knox County. I was twenty-
eight years old and had never been to Frankfort. I had never
been to Georgetown and had only been to Lexington five
times. The testing began at 8:00 a.m., and I left home for the
120-mile trip at 4:30 in the morning.

During the testing, I discovered that I was not as skilled
as other applicants on the assembly projects. I tried to make
up for this by being positive, trying very hard and encouraging teamwork during the team exercise. I was then called for
an interview. I was interviewed by Tom Blue, Bob Evans, and
Mike Stefanski in PC (production control). I was called in for
a physical shortly after that. I received some bad news during
my physical. I failed the colorblind screening test, and was told
to visit a doctor for a more detailed colorblind test and have
the doctor send the information to Toyota. I did not think I

would hear anything else and was very disappointed, even though I was now working full-time at my current job.

I had a five-year-old daughter and a sixteen-month-old son. A big change was on the horizon for all of us. I came home from the grocery one day to find my mother standing on her porch. She excitedly told me that Toyota had called: "They are going to call you back very soon." Since I didn't have a phone, I went to my parents' house and waited. John Owens called and offered me the position of team member in production control. The offer came on Good Friday. I really do not have the words to express what that meant to me then and what it means to me now.

I started to work on April 10, 1989. I worked as a team member until 1992. During this time, I learned to drive a forklift and unload trailers. I also learned to pack and export parts.

I then worked in the kanban room for two years. I learned the kanban system during this time. I became a team leader in 1992 and a group leader in 1994. I could have pursued other opportunities in the company, but being a group leader gives me the chance to work more closely with the team members. I've always understood that the team members are really what make us successful. Of course, the group leader supports the assistant manager and manager but must support the team members every day. I think the group leader should care for his team members and not just care for the work they perform. I still believe this is the way Toyota cares for us all. I have always been blessed with great assistant managers and managers. I am very happy working with my current management team.

I had never heard of workers being called "team members" until I came to Toyota. Now it seems to be common in a lot of workplaces. The very best thing about working at Toyota, for me, is very simple. For the past nineteen years and nine months, I have not had to decide if my wife or children were sick enough for medical attention. I have not had to wake up in the morning wondering how I would provide for my family. My wife has always supported me in my job even during the times when the hours were long. My family has always been proud of my position as a Toyota team member. I hope my Toyota family is as proud. Thank you for the opportunity.

April 10, 1989, was truly a Good Friday after all.

> I have grown to admire this man a great deal. He cares for all his team members on a personal level. Audie is the group leader responsible for making sure the parts make it to the lines. This is a very tough job. —Tim

"You're only successful as long as you never forget where you came from."

Doug Omohundro: Assistant Manager, Conveyance

Toyota … "What a Feeling"

Being employed by Toyota over the past nineteen years has been very rewarding in my own personal development. The fostering of self-worth is one of the most fundamental foundations in the Toyota business model. The self-worth is complemented by the team concept. Your ability to contribute to the team allows for development of self, which in turn improves overall morale. It is this high level of morale that drives our company to great success every year.

I make it a point each day to focus on the improvement of morale within the workplace. It is very important to make yourself available to the TMs whether it is work or nonwork related. I started off myself as a team member on the production floor. This allowed me to understand the importance of the interactive management approach. I have a certain phrase that I attribute to my own success, and that is "You're only successful as long as you never forget where you came from." Keeping that in mind will allow good decision making that will benefit both the company and the team members.

Kaizen is also one of the main elements to Toyota's success. It is not intended to ever create a comfort zone in which no one is ever motivated to change. It is the constant change that creates both individual and company growth. I enjoy being a part of a company that believes in its people so much that it allows us to make the changes within the workplace that will continue to place Toyota as a healthy company in such a volatile environment. Both my family and I are blessed to be associated with a company who looks out for our well-being today and in the future.

Yosh!!!!!!!!!

> Doug's job is as an assistant manager in our chassis section. Doug is always on the floor visiting with his team members. He makes it a point to go out and wish them a happy birthday. These small personal acts are what make Doug such an excellent leader. Many people as they get promoted and move up in a company forget where they came from. I believe that in Doug's case, this will never happen. —Tim

"I know you don't get a second chance in life, so I
said, 'You bet I would like another chance.'"

Lowry Gillis: Team Leader

I was hired on May 23, 1988. I previously had a job that paid
well in Louisville, Kentucky, but it went out of business so
of course I had to seek employment elsewhere. While working at another job in 1986, I heard that Toyota was coming to
Kentucky and it sounded like it would be a job with good pay
and benefits.

So, like many other people, I applied for a job and figured
with the amount of applications that I didn't have a chance. I got
called to go through all the testing that was required. On one
of my assessments scheduled in Frankfort, I was sick with the
flu and missed it. I thought my chance to work at Toyota was
gone. Around a month later, Toyota called and asked if I wanted
to reschedule my assessment. I know you don't get a second
chance in life, so I said, "You bet I would like another chance."

I was hired and started to work in bumper paint. The
only thing I knew about paint was how to roll it on a wall.
I received a lot of training and in my mind became a pretty
good painter. Then my opportunity came to broaden my
knowledge of paint even more. I was sent to Japan for training. I went to the PPG paint school and to the Nachi Robot
School. I am thankful for all the classes and training I
received. There are not a lot of jobs that give you a chance at
that kind of training.

Later on, I got the chance to transfer to what I thought
would be a little easier job for an old man like me in PC,
which eventually became known as assembly conveyance.
Around one and a half years ago, I was asked if I would be
interested in becoming a GPC (Global Production Center)
trainer. I thought about it and decided that I should take
advantage of all the training I could get.

My job is to train transfers, new hires, or temps that come to assembly conveyance. I train them how to properly drive our powered industrial equipment. During this training, I tell everyone how I have worked for many companies in my sixty-one years and I have never seen a company so dedicated to safety as Toyota. The company talks, preaches, and practices safety.

I am proud to say that I am a GPC trainer because it is the best training I have ever seen. It is a three-day training session. One hundred percent of the team members trained say they are very impressed with the training. I am proud to know that I am contributing to the development of our future team members.

> Lowry Gillis has worked in many areas of the plant, which made him a perfect choice to be a GPC trainer. His desire to teach is shown through the success of the team members he has taught. —Tim

"Let's take a ride."

David Craig: Assembly 1 Safety Team and Proud of It!!!

Why take a ride, you ask? Well, that's how my Toyota career started—by taking a ride—so allow me the pleasure of sharing my ride with you. Our journey begins in late spring of 1990 while taking a pleasure ride through the country. I had sat out the past semester of college on the false pretext that I had a full-time job offer at one of the major plants in my hometown. They decided not to hire me, stating that I needed more work experience or that I needed to finish college before they would hire me. That was nice to know now that I had decided to sit out the semester thinking I was hired.

So there I was taking the scenic route, passing the very place I thought I was supposed to be working at. Ironically, my drive takes me past an unemployment office in Florence, Kentucky. While I had no desire or right to draw unemployment, I decided to stop in and see what jobs were available. Little did I know my future was staring me in the face. There it was: a sign advertising that Toyota Motor Manufacturing in Georgetown, Kentucky, was hiring. I filled out my application, and from that day forward my life would change. Everyone said that it takes years for them to call you back and the waiting list is so long you'll never hear from them. So as usual I bought into what they were saying and went along with the crowd, not expecting a call.

I reentered college, which was a good thing, and took a job in my small hometown hoping that one of the local industries would decide to hire me. I eventually got a call to take the assessment for Toyota, and from that point on I was relentless. I would call weekly to see where I was in the hiring process. In what seemed like an eternity, I got the call for the medical assessment, and shortly after that I got the call that would change my life.

I will never forget that night, I was at my girlfriend's house when my dad came by and said, "You had a call from someone about a job." He couldn't remember who it was, but he had their number. I called, and a gentleman named Charles Duke answered the phone. When Mr. Duke offered me a job with Toyota, I could hardly catch my breath. I am not sure who was happier that day: me, my dad, or my girlfriend. See, there was more to this phone call than just another job offer. Only a week earlier, we had found out that my girlfriend was pregnant. We had no insurance and really no stable way to raise a family as we were both in school.

On February 25, 1991, I walked through the doors of Toyota for the second time, and this time I was one of them. It was as if I had known these people all of my life. I was immediately welcomed by those in charge of training the new hires, and after a week in the classroom I hit the floor. My classroom was eighty-five acres of concrete and steel. I would learn more over the next eighteen years than I could ever have imagined.

One may think, "Well, anyone can learn how to put parts on a car or drive a fork truck." Learning my process was the easiest thing I ever did. While learning my process was crucial to my success at Toyota, it was only the beginning of a continuous education. I remember hearing one of my early managers saying we build people here at Toyota and we also build cars. I cannot begin to express the way Toyota has molded me into the person I am today. In being a Toyota team member, we are expected to build only the highest quality car. The expectations we have are tremendous. Personally, those expectations that I have learned through the years have transferred into my personal life as well.

We learn at Toyota to do it the same way each and every time; this is called standardized work. Through this repetition, we become highly skilled in whatever our process may be. I carried this philosophy with me into the coaching of children in youth sports. By learning the correct method and applying it the same way each and every time and through continuous repetition of those skills, my players became very talented.

By applying the values I have learned while at TMMK, I feel I am a better all-around person because I have a higher expectation of myself as a person. Much like the cars we build, I have learned not to accept anything less than the best in all aspects of life. Does that mean I have to have the best of material things? No, not at all. What it does mean is that every day I get out of the bed is a new challenge to be the best that I can be in whatever I attempt on that day. Another quote that I have carried with me over the years is "Give a man a fish, feed him for a day. Teach a man to fish, feed him for a lifetime." I truly feel that the things I have learned have not only impacted my life but my children's as well, and generations to come.

> David Craig is my counterpart in the safety team. He is very dedicated to his work. I am sure he takes this dedication with him in whatever he does. I am glad to call him my friend because I learn from him every day. —Tim

Avery Bussell: Group Leader, Assembly 1

Farming—a lot of who I am and have become, I thank to my childhood of growing up on a tobacco and dairy farm in Fleming County, Kentucky. You may wonder why I talk about farming in a book about auto manufacturing. Farming is a perfect example of teamwork and problem solving in action. My plan was to be a farmer, but like so many others for financial reasons I had to look for other ways to make a living; so in 1987, at twenty-one years old, I applied for a job with TMMK. At the time, there were over 100,000 applications for only 3,000 jobs. I was called to Maysville, Kentucky, to do my initial testing. Toyota had committed to hiring its employees from all over the state, so testing was done in different cities in the region. Once I moved up in the testing, I was sent to Frankfort, Kentucky. My family was housing tobacco in late summer 1987 and we had stopped for lunch when the mail arrived, telling me that I was selected for an interview. It was a very exciting time for my family.

In February 1988, I did my interview to be hired. I was scared and intimidated at the thought of doing this interview.

I had some speaking skills but was still somewhat shy and nervous. I had been involved with FFA (Future Farmers of America) while in high school and was the president of that club. Little did I know what kind of impact being involved with FFA would have on my life. Cheryl Jones (former TMMK vice president) did the interview with me. I guess she could sense my nervousness, so she began to ask me what I had done in the past. I told her about my involvement with this club. She had a farming background herself, so she began to talk about farming. I owe a lot to Cheryl because of her people skills and her genuine belief in Toyota's "respect for people" philosophy. Cheryl gave scenarios from issues raised from farming and how I handled or would handle each of those issues. It would be years later before I would realize that she was interviewing my problem-solving skills. Fortunately for me, Cheryl would go on to be my group leader. As I stated before, farming would help me to be successful in a manufacturing setting, and Cheryl understood that from the beginning.

I spent nine months as a team member before being promoted to team leader in 1989. I worked on Trim 1. This is the first line in assembly after the cars are painted. Trim 1 takes off the doors and adds many wire harnesses that are located under the carpet and behind the engine. I would then be promoted to group leader and have been a group leader ever since. My leadership style is simple. I am only as successful as my team members and team leaders, and vice versa. I work for them. They do not work for me.

I had had some great opportunities to travel thanks to Toyota. I went to Japan in 1990. That was an awesome experience learning of a different culture. In 1995, I had the opportunity to work at the New York Auto Show for four days representing the company. I have also had the opportunity to attend and work at the last seven Future Farmers of America national conventions at the Toyota Booth and the career fairs. I was able to serve as a national judge each year.

Toyota sponsors the FFA, and I am very proud to represent both Toyota and the FFA in this capacity. FFA is a great youth organization that personally helped me to develop my interpersonal skills.

I have made hundreds of good friendships through my work, and my whole life surrounds this. Toyota definitely formed and shaped me into a responsible adult. The company has allowed me to live a wonderful prosperous life. My family and I are very thankful for the opportunities that have been granted to us through my work here.

> Avery is correct. Farming is a great example of teamwork, and it has had a positive influence on so many people within our plant in particular. —Tim

Chapter 3

Our Motivation: Stories of Appreciation and Dedication

In 1947, our company was in the bad situation of having to borrow money to maintain production. During that time, the bank decided that Toyota had too many workers and made the company lay off some employees. This was a low point in company history that caused a shift in thinking by the executives. They vowed at that time that no outside factors would ever play a role again in the decisions made by the company.

They paid back what money was owed and made plans to save money for "rainy days." Well, those rainy days are upon us. The company has held true to their values and kept us all working even at the loss of making money. The company, of course, cannot do this forever, but now they are holding true to their word.

The following stories are from team members with different backgrounds, but all with a similar message of appreciation and an increased dedication to the company that provides us all with steady work during this downturn in the economy. We are all learning and growing every day. Having this security allows us to focus on our families to better their lives. Our job security is just one of the factors that make us proud to come to work every day and build the best car we can.

"My journey has now come full circle. I've had the experiences of helping start a new plant, and today I am able to give back to the ones where it all began."

Shannon Conder: Team Member, TMMI Assembly Conveyance

I've been asked to tell my story ... the story of a Toyota team member. I would guess that mine is going to differ from others in a major way: I am a TMMI (Toyota Motor Manufacturing, Indiana) team member dispatched to TMMK.

Now, for you to fully understand, I need to take a few steps back. Just over three years ago, in December of 2005, I was asked to join the TMMTX (Toyota Motor Manufacturing, Texas) start-up pilot team for assembly conveyance. I was given all of the details and told to go home and take the weekend to think it over. I did not need one minute, much less a full weekend, to make that decision: I knew I wanted it! I wanted to grow and to learn, to teach and be taught, and with the invitation, I knew I would experience it all. Through great management and support from fellow team members, I had quite the journey. I felt I had found my purpose at Toyota. Each team member I taught in turn taught me about the type of team member (and person, for that matter) I wanted to be. That leg of my journey lasted nearly a year and a half, and I loved it!

Now to bring us closer to today, you also need to know that I returned to TMMI as a production team member. At my plant, we built the Tundra (now solely produced at TMMTX) and the Sequoia. With the economic downturn well into 2008, we began to feel the impact at TMMI. Every day brought a new set of worries to team members, but continuously we were reassured by our management to not worry ... our employment with Toyota was secure. Toyota operates with the philosophy of not laying off, that its employees are its greatest assets. So as unheard-of decisions were being made at the top (halting production at two plants, moving a new model into one plant and that plant's current model solely to another site), they were educating their team members at home through excellent training to become world-class leaders. Hours upon hours of classroom studies ... Toyota's history, TPS, diversity, and the list goes on. Even community involvement projects were being evaluated and carried out, then finally came the dispatch opportunities. Yet another tool Toyota has provided its team members with, to teach and be taught.

Therefore, instead of sending us home, they have given fellow team members and me another opportunity to grow. Here at TMMK, I am doing just that. My journey has now come full circle. I've had the experiences of helping start a new plant, and today I am able to give back to the ones where it all began.

To me, being a valuable, honest team member is not an expectation; it is a privilege. I am thankful every day for each new challenge I am handed.

Thank you, Toyota, for all you continue to do for my fellow team members and myself.

Shannon is one of the pleasant surprises I encountered while writing this book and one of the few people I met because of the book. Her story came at just the right time for me personally. After she moved back to Indiana, I have stayed in touch with her through Facebook and e-mail. She is a great friend. —Tim

"Decisions like this make it easy for me to come through the turnstiles every day and do the best that I can to give the same level of respect back, and build quality cars."

Bryan Litteral: Team Leader

I grew up in a family where hard work and dedication to your work were very important. My family also believed that hard work and dedication would also determine how well you would succeed in life. I graduated in 1997 and took a local factory job. I worked there for about two years. In May of 2000, I was hired into the Toyota family. This was probably the best opportunity anyone had ever given me. I knew that if I worked hard, that this opportunity would change my life.

The first eight weeks of work, I spent most of my time in classes. The classes were given to me to help develop me into a team member. All of this was new to me. I was expecting to go straight to work, maybe a little on-the-job training, but not all this. This experience was overwhelming. I knew that this company was serious about its team members and their ability to succeed and push the company forward.

I started in Plant 1 Assembly on Final 1. From day one on the line, everyone was so eager to help me and teach me what he or she knew. Not just about the jobs, but about teamwork, communication, and diversity. I worked on Final 1 as a team member for about six years. In the course of those six years, there was a lot of change. Different team members, different jobs, and different takt time, but one thing stayed the same: Toyota's belief that people truly are their best asset.

I was promoted to team leader in September of 2006. Once again, I went to three weeks of classes. Since then, I have worked hard to uphold our core values and to pass them on to each team member, contractor, vendor, or anyone I encounter working at Toyota.

Toyota is a very successful company that gives each team member respect and security of long-term employment. Team members all across North America are experiencing this right now with the economic downturn. The way the company is maintaining our workforce and providing work even though no cars are being built says a lot to me about how they appreciate our hard work and dedication to the company. Decisions like this make it easy for me to come through the turnstiles every day and do the best that I can to give the same level of respect back, and build quality cars for our customers.

> Bryan is one of the group safety representatives. Bryan always wants to learn and be active in safety initiatives. I appreciate his devotion to his work. The group he works in installs the doors onto the cars. —Tim

"The opportunities to improve and advance are also a way TMMK sets itself apart."

Jeff Hill: Team Leader

I grew up the son of a factory worker and farmer. I am also the nephew of eight other factory workers. I know a thing or two about the life and times of factory workers.

As a child, all I ever heard from my dad and other relatives were stories about working in factories, and they all ended with the same line: "Jeff, get your education so you can get a better job." Well, I did get my education, begrudgingly I might add, and thanks to TMMK I got a better job.

TMMK is a better job for lots of reasons. First and foremost is because we build a high-quality product that is in great demand. That fact, along with the diligent application of TPS principles by the management of TMMK from its very beginning, make the stories that I tell to my children, nieces, and nephews about the life of a factory worker far more enjoyable. My stories don't include the horrors of layoffs and strikes or laments about doing the same thing day after day. My stories include tales of being the number one selling car in America, and perfect attendance ceremonies where you can win a new car, and how TMMK and its employees do great things in the community.

One item I always mention is how much TMMK has helped improve the schools in this area. With its annual "in lieu of taxes" donation to the local school system, totaling around $30 million, the positive effect to the education of all children in this area, including my two daughters, is quite immeasurable. Toyota as a whole sets itself apart from other employers in many ways, but perhaps the most important is its commitment to the long-term employment stability of its employees.

This is a concept that is proving to be most reassuring in the current economic downturn because of the instability currently in the automotive marketplace. The opportunities to improve and advance are also a way TMMK sets itself apart. The Reach for the Stars Program, special projects activities, and quality circles are all great tools available to anyone who wants to improve themselves and their problem-solving skills and thereby promote kaizen (continuous improvement).

My personal education and professional advancements have benefited from these tools. But perhaps the best resource at TMMK is the experience and ingenuity of its team members. I have learned so much from so many people in my ten years.

I hope I can pass on what I have learned in my time here to the newest team members of TMMK, so their stories about the life and times of a factory worker will inspire the next generation to continue to make and enjoy TMMK as a great place to work.

I would like to thank everyone I have worked with, and learned from, and I thank TMMK for providing great story material.

> Jeff was one of my team members while I was on the production floor. Jeff has now moved to a team leader role, and I am very proud to say he was once one of my team members. It is always fun to see someone you have worked with become successful and move up in the company. —Tim

"I grew up in the town where the War on Poverty was officially declared, and to say the least a job with good benefits and good pay was completely unheard of."

Katrina Bevins: Team Member

When I first learned that I had a job at Toyota, I was excited. I grew up in the town where the War on Poverty was officially declared, and to say the least a job with good benefits and good pay was completely unheard of. Some people tried to convince me not to take the job. They said, "It's a factory full of men; they won't treat a young girl right." I had some people try to tell me I would be wasting all the time I spent in college because I would just be thrown on an assembly line and spend the rest of my life there. But, I just couldn't let all the negative talk strike fear in me and keep me from pursuing a new opportunity.

The first day I was here, I was surprised to find just how many other young women worked here. I was welcomed by almost everyone I met. Many of the ladies I have worked with have become my friends. They have helped during my wedding and while I was expecting my first child. More surprising is how well the men in the factory have been to me. They do not treat me differently like people had tried to make me believe they would. Everyone here has become my family away from home. We talk about our kids and attend church together. We are involved in community activities together and even sometimes just spend free time together having a little fun.

I have only been here a short time in comparison to some of my colleagues, but I have already been given the opportunity to move to a few different areas and get involved in many projects. I have found that many of the skills I learned in college have been put to good use. I have even been given the chance to learn a few new things as well.

I am so grateful that I am now in the position to help my family financially and feel pride in myself for a job well done at the same time. I have had so many opportunities come my way the past few years, and so many great people enter my life. I often think I am so glad that I did not let the fear of what might happen stop me from pursuing what has turned out to be one of the greatest opportunities I have ever received.

> Katrina, I am thankful that you didn't listen to all the negative talk about working in a factory. Our plant is better because you are a part of the team. —Tim

"I have had perfect attendance since 1991 and have been privileged to be able to see great entertainers from David Copperfield to Alabama."

David Farmer: Associate Staff

In 1991 I was working for a computer company in Lexington, Kentucky. I was in a dead-end job for a company that was getting ready to quit selling computers. In February, they told me that I was losing my job. I was really worried about being able to provide for my family. I had always been blessed in finding good jobs.

I was at a store in Georgetown when I came across a friend of mine. We began talking about how I had lost my job. He then told me about a really good company named Toyota that was in Georgetown. He said that I could get on through a temp agency and if I show them my work ethic, there would be a good chance that I could gain permanent employment.

I went to the temp agency that same day. I was then placed in body weld as a temporary assistant staff. I was elated because I had done research on Toyota and saw what a great company they were.

I started as a temporary in March 1991, and by August I had become a permanent team member. I realized that I was again blessed by working for one of the best companies in America. I was thrilled to see all of the benefits that Toyota provided their employees. One of the best was job security. They always said that it was Toyota's goal to provide job stability through good times and bad.

Another of the things Toyota tries to do is to provide morale-building activities. One of my favorite activities that Toyota has provided over the years has been the Perfect Attendance Ceremony. I have had perfect attendance since 1991 and have been privileged to be able to see great entertainers from David Copperfield to Alabama.

In the 1999 Attendance Ceremony, I was in attendance when Martina McBride was the featured act. My wife and I had a great time at the event just like the other attendance ceremonies that we had been able to see. However, something really special happened to me that night. I won a Toyota Avalon. I can't express what I was feeling that night because I was in shock.

The Masters of Ceremonies were Crook and Chase. After they called my employee number and name, my wife ran to the stage. They asked me what it was like to win a car, and I was so speechless that all I could do was let out a laugh.

After riding out in the Avalon, we went backstage for some more celebration. I was able to personally meet Martina McBride backstage. I was so excited to meet such a celebrity. What was weird was that she seemed so down to earth.

Throughout the years, I feel that I have been so blessed working for Toyota. Toyota is not a perfect company, but the leadership there really tries to do things in the best interest of the team member. Their ultimate goal is to provide a stable work environment for everyone, and I am thankful for my time here at Toyota. It is my goal to continue working here until it is time for me to retire. Thank you, Toyota!

> David is a good friend. I asked David to write about his experience with winning the car, and I appreciate his willingness to do so. —Tim

"Simply put, I was hooked for life. I knew this was the place for me."

Rick Corbin: Group Leader, Body Weld 1 CCR

I drove by the plant during construction on my way to work for about two years. I was working for a company in Georgetown that did contract machining. I worked as a CNC machinist and later as a quality control inspector. I really never considered trying to get an interview for two reasons. First, all we heard was how only the top 3% of applicants were being hired. I certainly felt I did not fit the criteria. Secondly, I really enjoyed what I was doing. But really I was scared to try—you know, fear of rejection.

In late 1988, I found out that our factory was going to close soon after the first of the year. One day, someone from either Toyota or the state came to talk to us about applying and the hiring process for Toyota. I distinctly remember someone asking about how many of us could expect to be hired. The person said about two or three. There were sixty of us there at the time. I think about fifteen or so made it, but I'm getting ahead of myself.

On December 23, 1988, I was one of six people still working at the factory in Georgetown. I was sent home for good that day. I was beginning to get over my fear and hoping for the best. I interviewed for either maintenance or tool and die and was not selected. I was then asked if I wanted to be placed in the production pool; of course, I said yes.

I was hired on May 15, 1989. I started in body weld, and again I was scared. The size of the place was enough to scare you. I was totally impressed and sold on Toyota right off the bat. Even though it was rigid in structure, the freedom to be heard, to share ideas, to show what one could do amazed me. Simply put, I was hooked for life. I knew this was the place for me. For the first time in my life, I had to work only one job to make ends meet. I could think about things I could not afford to have even considered in the past, things such as vacations and retiring some day. I cannot begin to tell you how much this company has and does mean to me and my family. I truly thank Toyota, and I thank God for Toyota.

> Rick's current assignment is a group leader in body weld. Body weld and stamping begin the process of building the cars. Rolls of steel come into the plant. That steel is stamped and made into body parts. Those parts are then welded together, flow through the paint shop, and then enter assembly. Thank you, Rick, for participating with this book. —Tim

"During my orientation, I was told how Toyota operated like a bridge."

David Eads: Team Leader

April 1998 was a good time in my life. TMMK hired me after a four-year waiting period and my Kentucky Wildcats had just won their seventh basketball championship (that's important around here). I enjoy painting. I am a fine fisherman and a decent storyteller. You can't be good at one without being good at the other.

I started work on our Chassis 2 line installing engines and exhaust systems. I then worked on our Final 1 line, where my team leader was Larry McCray. Anytime he had a new team member, he'll pull me into a discussion for their benefit. It goes something like this: "Dave, how was your quality when you worked for me?" I respond, "Perfect; I had no problems." Larry would then turn to the new team member: "See, that's what happens when you build vehicles by our Toyota production system. Build each car the same way, every time."

Quality. It brings customers to us. Quality. It's what brings them back. Quality. It leads our customers to recommend buying a Toyota to their friends.

Next I moved to our Trim 3 line. There I installed seatbelts, tail lights, and bumpers. From there, I transferred to our conveyance section. I'd sequence parts such as radiators and deliver them to our line side groups for installation. I'm very fortunate to have learned and experienced much in our assembly shop. Toyota has always offered avenues to further one's education and abilities. So I participated in classes and was promoted to team leader on our instrument panel buildup line. That is where I currently am, but let's go back to my beginning at TMMK.

During my orientation, I was told how Toyota operated like a bridge. Our suppliers, materials, and parts were on one side of the bridge. Our customers were waiting on the other side of the bridge. The team members would make up the top portion of the bridge. They would manufacture the vehicles and move it to the customers. The supports of the bridge in our visual diagram were made up of team leaders, specialists, and management of all levels. This bridge, however, looks like a pyramid set on its top. That top piece supporting all the others is held by our president. These positions, these everyday people, would support the needs of those above them.

These positions would respond to our inputs on quality by contacting suppliers and engineering to address our concerns. They would help us make our jobs more efficient, safer, and more cost effective. They would encourage our ideas to simplify our work by reducing waste, saving energy, and protecting the environment. They would help us see our designs implemented.

The bridge works in other ways as well; while it helps us maintain building high-quality vehicles, it ensures we take care of our team members. Our everyday people support us during illness, family crisis, and house fires. During trying times, they are our good neighbors.

My trying time came in the form of a workplace injury. Putting 550 engines into vehicles every day is physically very demanding. Surgery was necessary. This was a very difficult time in my life. I had a young family to provide for and was nervous. If you've ever read the "Footprints in the sand" poem by Carolyn Joyce Carty, this was a time there was only one set of footprints for me.

I was focused on what was wrong, what needed to be done to fix it, and how long it would take to recover. I wasn't alone. My teammates always encouraged me. Such a fine group of men and women, never a sour or down word did they utter. My team leaders, group leader, and assistant manager always supported me. They would ask to bring food over, cut my grass, and change my oil. Someone always calling to check on me; they were really good people.

I feel Toyota's drive to build the best-quality vehicle also extends to providing the best-quality care for its team members. Our TMMK values are built on "Respect for people and our best resources are our people." I'm proud to work for a company that stands behind its employees. My success story is proof of TMMK's commitment to its people.

> Dave was promoted to team leader while I have been in Assembly 1 safety. It has been a pleasure to watch him develop as a leader. He has a very bright future at TMMK. —Tim

Chapter 4

Our Inspiration: Stories of Support and Triumph

I am sure each workplace has employees that inspire others through their struggles and how they respond during those difficult times. This chapter is about some of those inspiring team members at TMMK. These team members have gone through some difficult times, and along the way have discovered on a much more personal level how the company and their friends will pull together to support them.

We have a wonderful outreach program that helps team members in times of need. I can recall one story where a team member lost his father. He had to take his family to West Virginia for a week to deal with the loss and take care of the estate. Our outreach program provided this team member with financial assistance by helping his large family stay in a hotel. There are thousands of stories like that around our plant. When times get tough, they are there to help us.

"Outreach is your best friend if you ever find yourself
in this situation. I know from experience."

Larry McRae: Team Leader

I started work at TMMK in October 1995. I didn't know very
much about factory work, but it didn't take very long to find out.

My first six months were very much like basic training. I
was doing everything I needed to do in order to get my probation period over with and become a full-time employee.

After learning several jobs on Final 1, I was beginning
to think about the next level. My group leader at the time
thought I might make a very good team leader, so I took the
plunge.

After being promoted to team leader, I got a chance to
see the big picture. I was invited to round table meetings and
had the chance to talk with upper management and became
familiar with how Toyota operated.

The Boss (Bruce Springsteen) said if you pay a person a
decent wage, then you don't have to worry about a contract or
having a union. These are a few things that Toyota does that
set them apart from some companies.

First and foremost, we are treated with respect by our
supervisors and our peers. Second, you make a very, very
good salary. Third, they give you all the perks, bonus, vacation, company picnic, car giveaway, and don't forget about the
Outreach Program!

The Outreach Program is Toyota's best-kept secret. We all
hope we never have a medical, health, or a house fire where
everything is lost and you have no place to go. Outreach is
your best friend if you ever find yourself in this situation. I
know from experience.

I had esophagus cancer in July 31, 2003, and had to have
radiation treatments for the next three months. Outreach was
at my house every day, waiting in the driveway to take me

for my treatments. Also, while out on medical leave, Outreach gave me a gift card. I can't thank Toyota and Outreach enough for what they did for me during my illness.

I've made two very smart decisions since the age of eighteen.

One, I joined the Army, and the second is when I came to TMMK. I'm very proud to say that I work for the best company in the world.

> Larry retired while this book was being completed. He is a well-respected team leader who will be missed at TMMK. His legacy of hard work and dedication will continue through the team members he has led. —Tim

"At that moment, I knew that this wasn't just a good-paying job; it was something much more. It was a family."

Matt Lucas: Team Member

This is my story of how Toyota has blessed me and my family. I was blessed to have received a job as a temporary in August 2004. I was thankful to have a job that paid well. Even as a temp, I started out at close to $14.00 an hour. With a one-year-old baby girl, I was thankful.

Well, in April 2005 I wasn't feeling myself: very tired feeling and drained of all my energy. I really wasn't prepared for the news that followed the next few days. I had went to the doctor for a checkup for fatigue and returned to work that night feeling a little exhausted but dedicated to Toyota for the opportunity that they gave me. The next morning, I received a call from my doctor saying that I needed to report to the hospital immediately. That got me nervous and shook up. Well, when I got to the hospital they told me that I was in kidney failure. I felt like my life was a sheet of paper that had been crumbled up and ready to throw away. It was a hard pill to swallow, but the team members at work kept me in good spirits.

At that time, I was a temporary; the time that I would be off work, I would not be paid. All of the team members that I work with and surrounding lines came together and helped my family and me financially. I didn't know how to react. These people, that I had only known for a short period of time, would do this for me? At that moment, I knew that this wasn't just a good-paying job; it was something much more. It was a family.

I was out of work for only a month. When I returned to work, I was now taking dialysis treatments three times a week, and it was very hard to do that and work forty-plus hours a week. My doctors just could not believe that I was working and doing dialysis. They said that I was a very special person. That wasn't the case at all. The truth in the matter is that working at Toyota through all of this gave me hope and made me stronger spiritually. I continued to do treatments and work there for about ten months. In February 2007, I received a kidney transplant that was donated by my loving wife. Last year, we had a baby boy.

Through all of this, Toyota has helped me in more ways than one. For example, I am on some medicine that is very expensive, and the insurance alone has saved me thousands of dollars. The support of my fellow team members has made me feel as if this is my home away from home.

Through this whole experience, I can say that I have been very blessed because I have a really good job and, most important, I have made some truly good friends. My team leader was there for me and helped my family and me at a time when I needed help, and for that we are truly grateful.

Our president at Toyota is a very caring person because he has been there and knows what it is like on the factory floor. He understands the work and the purpose of his position.

I think that Toyota is very lucky to have people like this working for them. When times get hard, they are the first people to stand up and do something about it. Like I said, Toyota is so much more than a good job,

We are a family.

> Matt, you have inspired many with the way you handled yourself as you went through your tough times. I am proud to know you and call you a friend. —Tim

"After this, I vowed to give Toyota 100% in every aspect of work I could do and to learn as much as I could—as I call it, 'pay them back'—for all the help they have given me."

Brent Pennington: Team Member

My story begins in 1998, when I was blessed with a job at Toyota of Georgetown, Kentucky. My first assignment was in the final zone, or what we called at the time "the money line" because this was the last line before the vehicles were sold and out the door. I saw what kind of effort went into assuring quality to our customers from my experiences on Final 2. Even the smallest discrepancy was repaired; sometimes they were so small you had to look twice before you could actually see them. This amazed me with the kind of detail they went through to make sure the customer received a perfect vehicle.

I saw how each team member was allowed to work independently on their processes after the adequate training that was required. This, to me, gave a feeling that you were needed and they were depending on you. As everyone knows, it is good to feel needed.

During my first few months of employment, I had a tragic event happen when my home caught on fire and burnt down. My wife and two kids were fine, but we lost everything. After this, I had to call Toyota to let them know what had happened and that I wouldn't be able to report for work for a few days. I was expecting the worst because Toyota is very big on attendance, but to my surprise they were very understanding and willing to help me in any way they possibly could.

When I returned to work five days later, I was called to my group break area to talk with my group leader. He told me about the benevolence fund that helps team members that have had a tragic event happen in their lives. I thought to myself this was great, and I wondered what other company would do this. Toyota helped me with money to get back on my feet and also helped with buying all new appliances for my new home. I was still in my six-month probation period, and to receive this kind of help was unheard of in the community I live in.

After this, I vowed to give Toyota 100% in every aspect of work I could do and to learn as much as I could—as I call it, "pay them back"—for all the help they have given me. Now it has been over ten years since I walked through the doors at Toyota; I have seen a lot of changes in my time here. I have seen how Toyota really cares about their workforce and what lengths they are willing to go to assure a stable work environment for each team member.

Toyota is a very diverse company and encourages the team members to be creative in helping the company to improve. This makes me very happy to work for Toyota of Georgetown, Kentucky. With the economy in the shape it is in, everything that Toyota does makes me proud to be a part of such a company, and I truly believe that if other automobile manufacturers were as good as Toyota they wouldn't be in the shape they are in now.

> Brent works with me in safety and trained me in my job. He is my mentor and friend, and I am grateful for him. He is also a dedicated father and just a great human being. —Tim

> "'Change is constant' is what I remember hearing from
> a veteran team member; I didn't understand what he
> meant, but within a few months I realized the message
> he was trying to get across was 'Always improving.'"

Nick Ousley: Team Member

If I told you that I love my job, would you believe me? If I told
you that I take great pride in my job, would you believe me?
And if I told you at the end of each day I am satisfied to know
that I've done the best I can do, would you believe me? You
should!!! It's true. I have always wanted somewhere to work
that wanted me as much as I needed them.... It took some
time to realize, but I found my calling at Toyota!

I started my career at Toyota Manufacturing in 2003 and
wasn't sure if this was the place for me! Cars everywhere! I
couldn't look in any direction without seeing a half-assembled
Camry charging down the assembly line, people in a hurry,
robots working nonstop, and a tour of visitors watching us
work every day! That's a lot of stuff for a new guy to process....

"Change is constant" is what I remember hearing from a
veteran team member; I didn't understand what he meant, but
within a few months I realized the message he was trying to
get across was "Always improving." This is the backbone of
Toyota: find a way to make something work, then improve on
it every day! This way of thinking keeps everyone's creativity
and imagination on edge by trying to have a brilliant idea
tomorrow from the outstanding invention yesterday!

Toyota had given me plenty of challenges in the short
time of my employment, but my task of knowing who I really
worked for was yet undetermined; let me explain....

I had taken a few days off work in the summer to have
fun at the lake, fun in the sun! A small vacation with family and friends just for the week grilling out burgers and hot
dogs, listening to music extremely loud, checking out the

bikinis!!! Nice! It was probably one of the best-planned vacations. Friends with a houseboat, beds to sleep in at night, and a small boat, and Jet Ski to run supplies.

Nothing ever goes as planned; a couple of guys couldn't get off work that evening and asked if I would pick them up early the next morning at the marina, and of course I did. That Saturday morning I woke up at about dawn, took a shower, got dressed, and went to untie the boat, but decided not to take it—just too loud! I didn't want to wake everyone up, so I decided I would make two trips to the marina on the Jet Ski. I untied the Jet Ski, gave a hard shove, and here I come fellas! The sun was up, beautiful and warm, when I idled into the marina and picked up passenger number one. I zipped back to the party barge, and away I went for passenger number two. I hadn't seen anyone else on the water that morning, and was about halfway to the marina when I spotted someone. To my surprise, he was on the wrong side of the lake; thinking quickly, I avoided a collision, but I hit his wake and plunged into the water...!

Disoriented and in pain, I swam to the Jet Ski and pulled myself back onboard, but I couldn't set down! Every attempt to set down was met with agonizing pain, I knew then I had a serious injury! To make things even worse, my means of transportation were damaged and had a top speed of 10 mph! A minute felt like an hour, but I finally made it to the marina, then it was straight to the emergency room.

Here comes the bad news: a crushed vertebra.... When people say their lives flash before their eyes, believe it, it happened to me. I said, "What do I do now?" In an instant, everything I'd worked for is gone, my life had changed.... So many questions with very few answers as family, friends, and surgeons tried to get me ready for the future.

Then came the hard part: I had to call my group leader and tell him that I may not be back to work. With that phone call, being sincere, he assured me that things are not as bad as they seem. He told me that my job would be waiting for me when I was ready to come back. That inspired me in a way that's hard to put into words, knowing that I'm not just a number in a computer. Toyota was at my side the whole time I had surgery and four months of therapy. Someone was always checking on me, asking what I needed or if they could help and wanting to help. Toyota's insurance had covered all my medical expenses, partial pay while rehabilitating, and still had my job waiting on me! To me, that was when I realized who I worked for: TOYOTA....

This is why I take great pride in what I do and who I do it for; Toyota provided me a second chance at life. I'm thankful to be a team member at Toyota, and I try every day to think of some way to repay the favor. Thanks, Toyota.

> I remember one Friday night, a group of us stayed and worked on some safety projects. He worked until 7:00 a.m. on Saturday morning and never complained once. You are a good man. Nick works on the door line. He was promoted last year, and I know he will do a great job. —Tim

"For several weeks, I kept getting envelopes in my locker, some with money, but all of them had words of encouragement."

Greg Richards: Team Member

It is very exciting to be a part of something so positive and optimistic at a time when it would be easy to be pessimistic and negative.

My stay at TMMK has been a great experience. It began nine and a half years ago. The idea was always that if you got hired at Toyota, all your problems were solved. Well, I am still waiting on that to manifest because it hasn't happened yet. The truth is that Toyota is simply a factory that builds cars, trucks, vans, and other types of machinery. It's the people who work at Toyota that make it a great place to work, and that is what I want to talk about.

I am aware of several Bible studies at Toyota that are conducted during our lunch breaks and have personal knowledge of managers, assistant managers, group leaders, and team leaders who support and attend some of these studies. Now in these times of political correctness, it would be so easy to simply say no to such groups meeting in the workplace. Toyota allows people of all religions to meet and fellowship with each other. The company sees the importance of people who want to reach out to other people. That is why this book is so important. This isn't a book created by Toyota management. It's a book created by team members to be able to share our values with the world.

The stories that are shared in this book are stories of optimism. I hope my story gives someone insight into the fact that people care for other people.

As a temporary, I was having a terrible day. I hadn't been at Toyota for nine months when my wife and I had one of the worst scares in our lives. Our eleven-month-old daughter, Kaleigh, was diagnosed with Type 1 diabetes; she had to stay one week at a children's hospital. It was a very trying time in our lives. I never thought that anyone at Toyota really knew me, or my situation with my daughter. After I returned to work, a stranger came up to me and handed me an envelope. He said it was taken up at a Bible study where someone had requested prayer for my family. The envelope had close to $1,000 in it.

I began to cry. He hugged me and told me, "See, people do care." For several weeks, I kept getting envelopes in my locker, some with money, but all of them had words of encouragement.

I believe in these times it is us, every human being, who need to circle the wagons and understand that we are all in this together. We may not see everything eye to eye. We are still neighbors. We all have a common goal that we all succeed. The best way for that to happen is for people to start helping people in their time of need.

> Greg is one of my door line friends. He is also one of the most faithful people I know. His faith in God is inspiring to many. —Tim

"My fellow team members collected money, and with the help from the Toyota Benevolent Fund bought supplies and showed up at our house to build a wheelchair ramp so my wife could come home. This ramp was a thing of beauty."

David Soard: Team Leader

It was September 9, 2004. At the time, I was a team leader on Chassis 5 in Assembly Plant 2. One of my team members was off that evening, so I was on line at left-hand ground tire. One of my fellow team leaders, Mike Thomas, came up to my process. He seemed to be a little fatigued as if he had run all the way up the line. I knew I didn't have an andon on and was unaware if the team member across from me had pulled his. But Mike didn't cross the line to the other team member. He stepped into the process and informed me I needed to call security dispatch at 4444. My heart sunk. Some folks will get emergency phone calls for the slightest problem at home. I, however, had never gotten an emergency phone call in the five and a half years I had worked here.

I walked over to the nearby phone on the tire platform and called 4444. I told the dispatcher who I was and that I had been told to call. "Hold one moment and I will connect you to the hospital rep." I didn't think my heart could sink any lower, I was mistaken. A nurse came on the line and asked me if I had a daughter named Sarah Soard. Sarah Ann is my pride and joy; at four years old, she is already showing signs she will be smarter and a more thoughtful person than her father would ever be. I could only say, "Yes." "I am the ER nurse at the Clark Regional Medical Center; we have Sarah in our ER. She is OK; she is just scared and needs her father." I don't know which was racing faster at this time, my heart or my mind. "What is she doing there? Where is her mother or my other two kids?"

"Sir, I do not know where her mother or your other children are. All I can tell you is Sarah was in an automobile accident and they brought her here. We were able to get your name from the car's registration and look up your name in our computer." I once again asked the lady, "Where is the rest of my family? Sarah couldn't have been in the car all by herself." She couldn't give me any other information. I yelled to Mike as I ran past, "My family has been in an accident and I am heading home."

From the moment I was heading toward the door, I was on my cell phone, calling friends and family to get and give any information I could. No one knew of the accident prior to me calling. As I got in my car, I called my best friend, Ronda, and asked her to get to the hospital and be with Sarah Ann. I called the Winchester Police Department and asked if there were any accidents. The dispatcher told me they were not working any accidents at this time. None of this was making any sense. Where could my wife be? I tried her cell phone; no answer. As I am traveling on the interstate, Ronda calls me back and tells me she is at the hospital and is with Sarah. Living only two blocks away is definitely a blessing. She also knows my son, Lucas, is OK. He is at baseball practice, and her sister has him with her. Well, that only leaves my wife Darlene and my eldest daughter, Ashley, missing. As I get closer to Lexington, I have to make a decision. More than likely, they would have taken Darlene and Ashley to a Lexington hospital. Do I head toward Winchester and wait for word, or do I go to Lexington and start checking the area hospitals? I call Ronda and ask how Sarah is doing. "She seems to be fine, but I haven't talked with a doctor yet." Knowing Sarah seems to have calmed down with Ronda by her side, I decide to head to University of Kentucky hospital and see if they have Darlene and Ashley. As I am pulling into the garage, I get a call from my sister-in-law. She went to our house and found Ashley at home asleep. She wasn't feeling well and stayed home while Mom and Sarah left to pick up Lucas from baseball practice. Now the only unknown is my wife. As I get out of the car, the phone rings again. My mother tells me Darlene is at UK, and it doesn't look good.

I get inside and go to the desk and ask about my wife. They tell me she is there and I can go back after they get her stable. In the meantime, they want me to register her. As I am giving them information, they start asking about Sarah Ann. "Why do you need her information?" I ask. "They are transporting her to UK as we speak, by helicopter," I can't believe what he is telling me. "I just talked with my friend, who is with her." I say. "They told me she was OK. What is wrong with her?" "Sir, I do not know; I just know they are transporting her here as we speak." I look at my phone and see where Ronda had called, so I call her back. "They decided to bring her to UK." Ronda says. "She has a seat belt injury around her waist, and they want to take her to the Children's Hospital in case she may have any internal injuries."

This night was the longest and most painful of my life. My wife and youngest daughter were in a collision just outside of Winchester. A teenage boy takes a curve too fast and comes on my wife's side of the road and hits my wife's car head on. Sarah is lucky. She looks worse for the wear. She is bruised and scared, but otherwise uninjured. My wife, however, had both legs broken, her pelvic bone broken, and her right knee is shattered in about a hundred pieces. She was in the hospital for two weeks and was transferred to the Winchester Center for Health and Rehabilitation, where she would stay until the day before Thanksgiving.

So what does all this have to do with how I perceive working at Toyota? Well, I loved my job before the accident, and I still love working here today. But after our ordeal, I learned about how caring my fellow team members are and how well Toyota takes care of its team members in times of crisis. My fellow team members called, visited, and showed support while I was off work. Toyota allowed me to take as much time as I needed to be with my family and to get our lives back in order. Toyota worked with me on scheduling conflicts so that I might return to work and still take care of my wife. An assistant manager made arrangements for me to work day shift so that I might take care of my kids and other family matters. My fellow team members collected money, and with the help from the Toyota Benevolent Fund bought supplies and showed up at our house to build a wheelchair ramp so my wife could come home. This ramp was a thing of beauty. It was as solid and good-looking as any deck I had seen. These team members gave their money and a weekend of their time to make my wife's transition home an easier one. Our health insurance policy through Toyota covered every medical bill. Not one time did I ever have to deal with a call from a representative of a hospital or doctor asking for payments. We could simply concentrate on my wife healing.

Everything from the Benevolent Fund sending me vouchers for food at the hospital to many, many people stopping me and asking about my wife and daughter every day proved to me that I worked for a truly remarkable company. I sometimes have complaints about my little world at Toyota, but I can tell you I know not of a company that would have cared for me more than Toyota did then. Thank you to my fellow team members and Toyota for making our lives a little less painful during this ordeal.

> Dave, I remember the night you left after your wife's accident. Everyone across the shop was praying and hoping that everything would be all right. I am glad to see that everyone's prayers were answered. —Tim

"The following three years were tough. My three teenagers and I did whatever we could to pay bills and put food on the table."

Kathy Walker Taulbee: Team Member

Toyota changed my life....

I married when I was eighteen years old. My husband had a job that provided for our family quite well. We were married for twenty-one years, and then he left. I had three children to raise on my own without his help or child support. Our home had to be sold along with our farm animals.

Not knowing what I was going to do, a friend (Betty) took me to the unemployment office in Georgetown to apply for a job at TMMK in 1993. I had never worked in a factory before. The following three years were tough. My three teenagers and I did whatever we could to pay bills and put food on the table.

On December 6, 1996, Toyota hired me!!!

In April 2000, my mother passed away. The company sent flowers and was very supportive. My group leader (Clarence Smith), God love him, hugged me that night I got the phone call. He was there for me. He even walked me to my truck that night and asked me if I needed any money.

Well, my life has changed for the best. I've seen my three children, Chad, Mississippi, and Jessie, grow up, finish high school, and have beautiful weddings. They have given me FIVE beautiful grandchildren.

I have also made great friends named Andy and Dawn Walker who work at TMMK. We camp and ride horses in the summer. I have also been able to see places I probably wouldn't have been able to afford. I have been to Hawaii, Australia, Alaska, Mexico, and I even ventured out west.

I still work very hard. I have good days and bad days. I am now fifty-five years old but can still keep up with the young kids.

Thanks, Toyota!!!

> I remember when Kathy was hired at TMMK. I was fairly new myself but was given the opportunity to train her. She always rose to the challenge. —Tim

Chapter 5

Our Determination: Stories of Kaizen

The term "kaizen" means "continuous improvement." It is common language at our plant. Everyone is looking for ways to improve our processes, workplace, and lives. Everyone wonders what the "magic" of Toyota facilities is. It is simply that we are never resting. We are constantly looking at ways to improve. It is embedded in us from day one to always strive for a better way.

This chapter focuses on some of the support members who promote kaizen and also some examples of improvements that have happened within our assembly shop. This constant thinking is the way we can build such a high-quality product and keep it at a similar retail price as our competitors'.

In some cases, the kaizens were done to provide a safer work environment for our fellow team members. In other cases, they were done to improve our quality. In every case, it was to add value to our company and products.

"We are all in this together."

Kelly C. Cox: Engineer, Power Train

Before coming to TMMK, I was an electrical engineer and project manager for an international material-handling company. This allowed me to work with many Fortune 500 companies for seven years. The only reason I mentioned my work history is that I have work experience to compare Toyota philosophies versus other companies'.

Currently I am in production engineering and responsible for line support of machining and assembly lines, along with leading and managing medium- to large-scale projects. We are finishing up a project (the AWD AZ Corolla Engine Project) budgeted for $712,000. Even before the economy started going downhill, we (power train, or PWT) were, of course, in the normal mode of cost savings per Toyota's general kaizen philosophies. The Corolla Project allowed TMMK power train to continue producing the AZ Engine (an original forecast plan of 6,150 engines per month). This engine will be shipped to Nummi, California, and also exported to Canada. Previously, years ago, we had similar export business (to

TMC [Toyota Motor Corporation]), and this method was what the TEMA Ringi of $712,000 was based upon to modify our assembly lines to manufacture this AZ engine. However, after I started the initial investigation of proceeding with this project, I noticed several areas in which we could improve efficiency (less conveyance), eliminate processes, reduce assets, reuse obsolete equipment from the vehicle plant (pneumatic tilt tables, flowracks, fork truck stops, etc.), and reuse excess equipment in power train (hoist, jib cranes, flowracks, tools, tool gun and controllers, etc.).

Now this sounds exactly like what a production engineer should be doing anyway. Right? Well, I agree.

But the cost-savings ideas I had were pretty radical and a big change from the way we had previously conducted business. When I explained my ideas to upper management, at no time was I confronted with the usual:

NO, we have always done it this way and there is no need to change.
Well, we will let you do it, but if it fails, it's all on you.

Instead, upper management creates the atmosphere of "This is a good opportunity for us (TMMK power train) to learn and grow and continue to be self-sufficient." I am given any and all resources to successfully complete the project ahead of schedule and under budget. With this type of support from upper management, it fosters the engineer's attitude of "FAILURE IS NOT AN OPTION." The result was a 48% reduction of project cost based upon the original budget, saving TMMK $398,000.

That alone would be reason to judge the project a success. But the story does not end there. With the economy progressively declining and our E/G volume sliding from 10,000 engines per month to 6,150 engines per month, we were able to reinvest a portion of the savings of $398,000 to purchase a machining station to add to our existing crankcase machining line that would perform the AWD (all wheel drive) clearance cut for additional Matrix, Vibe, and Corolla business that we (TMMK power train) were not previously considered for. This resulted in an addition 1,845 engines per month, helping keep our TMs (team members) employed during bad economic times.

What I take away from this experience compared to other companies is Toyota:

Gives individuals the support to be successful with the understanding of "FAILURE IS NOT AN OPTION."

This support is "THE RULE—NOT THE EXCEPTION."

Toyota (PWT), even after nineteen years, continues to learn and grow by accepting new and different ideas, not becoming a stagnant company.

But yet most effective is that upper management's leadership is based upon "LEADING BY EXAMPLE" and "RESPECT" versus other companies' methods.

This fosters TMs to take ownerships in their jobs and helps us realize:

"We are all in this together, One team on all levels."

I had asked Kelly to give an example of one of his projects. I appreciate you doing your part to improve our plant and helping keep us working during these hard times. This story is a great example of thinking outside of the box to solve an issue, and it greatly improved our company. —Tim

"When safety issues are eliminated and morale is elevated, this motivates me to keep improving every process and situation that I have the opportunity to be involved with."

Brian McElroy: Team Leader

Since joining Toyota in April of 1998, I have discovered that you must have an open and positive outlook on every aspect of life: from growing up in a small town to being part of a 7,000-person team, consisting of many different people with very different ideas, concerns, physical needs, goals, and language barriers.

I have learned people have many ways of communicating. Each person has a different temperament, which creates the need for each conversation to be exchanged in a respectful, horizontal level. If you are talking down to someone, you are wasting your time because they are not going to absorb anything you are saying. A sure way to fail is, while someone is talking to you, to be concentrating on your reply about how they are wrong. Each person has something to offer; we must learn to detect, understand, and utilize his or her contribution.

I believe there is more than one answer to each and every problem. Sometimes the larger task is to realize that there are better ways to do a process even when we do not see an issue. This is a great example of why we need to talk to the team members. They have a wealth of knowledge about what's happening on their processes.

For several years, I have been part of a safety kaizen team. We started on very small projects; this caused us to think about problems in an entirely different way. We had to find safe, effective, and efficient answers to each situation. While doing these kaizens, we had to learn who, what, and where our resources were. We found people who were willing to help us.

Communication

In order for the team members and Toyota as a whole to be successful, we have to share our knowledge. When we do this, all involved will prosper. Our safety kaizen team now has team members join in on some of our projects. This way, we can pass on what we have learned over the years, and we can also learn from them. Each kaizen we do, we see better ways to solve the issues our teammates are dealing with. When safety issues are eliminated and morale is elevated, this motivates me to keep improving every process and situation that I have the opportunity to be involved with. I feel our safety kaizen team's success has a lot to do with the freedom we have been given to be creative and make things happen.

> Brian can solve any problem and do it in a simple way. He has the rare ability of engaging the team members to be a part of the solution. It is a privilege to work with him. Currently he works on Chassis 2. They install many parts such as the front suspension. —Tim

"To me, that epitomizes teamwork. Selfless participation so the group can succeed."

Doug Collins: Team Leader, Body Weld 1

I started working at TMMK in April of 1997. Of my previous employers, the closest to Toyota was a supplier to Honda, where I learned about TPS and lean manufacturing, or their version of it, and kaizen. I had also spent many years in the entertainment business, several of that in a touring band, so teamwork has been essential in everything I've participated in.

I guess what stands out to me the most is how teamwork is a common thread in most successful ventures. Here at TMMK, not only do we participate in a team within a group, but also a team of groups within a shop and a team of shops within a plant, in addition to the many teams of administration and logistics that make this place run. So actually, we're participating in a team of several thousand people each day!

In my time here, I've had many good and several … challenging … experiences with teamwork. Regardless of which, each of them has taught me something positive about working with people and the importance of strengthening the framework of our teams. I have a couple of recent experiences I would like to share.

In May of 2008, I was asked to participate in a takt time increase project in Body Weld 1 to prepare for the lower volume in August. That team was made up of an AM (assistant manager), a couple of GLs (group leaders), and several TLs (team leaders) and TMs. We were given a scope of work and a time frame that was rather aggressive. The group represented either the front side of the shop, shell body and final line, or the back side of the shop, underbody, side member, and FB. We broke out into smaller teams within that team to tackle the issues. All the standard work had to be rewritten for the entire shop. An enormous amount of data had to be collected, analyzed, and put into form in a relatively short amount of time. This was accomplished through the efforts of many people working as a unit.

It was amazing to work with that many talented people, who put egos aside, realized each others' strengths, and deferred to those strengths so we could work together to accomplish the task. We had to work past group, shift, and even some shop boundaries to meet the deadlines and requirements. We were able to present a working suggestion by the time needed. Was it perfect? ... No.

Was it a good effort and example of teamwork? ... Absolutely.

After the first of this year, 2009, it became apparent that we were going to have to make some more serious and unpleasant decisions and increase takt time again, slowing down more. I volunteered to leave my group and be part of a project team, not really knowing what it would entail. I ended up being assigned, along with another TL that I knew but had not worked with, and several TMs to take over what had been a contractor's job of loading parts on AGV dollies for transport to the side member line.

Being from underbody, I wasn't familiar with these parts. There are two different load areas for these parts. Chad, the other TL, was from side member, so he took the more complex area and I took the other. Our task was to learn the jobs in two weeks and be able to take over. Because of our training and experiences with TPS and kaizen, we immediately realized there were several problems in the area that had to be addressed. Each of us, TLs and TMs on second shift and first shift, used problem countermeasure sheets and started listing the items that needed to be addressed.

Starting with the safety issues, we then began addressing one after the other. As it was last summer, people would recognize others' strengths and let each other work on things within their skill level. As a group, we were able to countermeasure several issues that were causing the contractors problems and improve the workability of the areas. With better labeling, visual guides, and standard work updates, we were able to make the areas much more efficient. First shift and second shift have been communicating well, and we will continue to improve the area. Again, this couldn't have been done without teamwork across group and shift boundaries.

One of my favorite quotes, though I can't remember who said it, is "It's amazing what you can accomplish when no one cares who gets the credit." To me, that epitomizes teamwork. Selfless participation so the group can succeed.

> Doug works in the body weld department. This is a great story of teamwork and what can be accomplished through good communication and teamwork. —Tim

"My group leader approached me one night and
asked me to try to provide a solution to a problem
on line."

David Foster: Team Member

You have probably heard the saying "Things happen when you
least expect it." This describes the first five years of my career
at Toyota Motor Manufacturing in Georgetown, Kentucky.

My story begins in late spring 2004. I was seeking new job
opportunities when I came across an advertisement in the
newspaper for production team member positions at TMMK.
The position was for temporary work, then moving up to
permanent status. At first, I decided not to apply because I
didn't want a temporary position, but then I came across the
same ad four months later and began to debate again whether
or not to apply. I finally submitted my application and began
working in a matter of weeks. My father worked in the auto
industry also, and I knew what the financial rewards could be.
My plan was to show my good work ethic and hopefully get a
permanent position.

My first days at Toyota were a bit of a shock for me. I knew
that the work would be physically challenging on the assembly line but was shocked to see how involved Toyota asked its
team members to be. Every job I have had in the past relied
only on the management to solve problems the company
faces, but Toyota was different. Toyota has a philosophy that
made so much sense to me. They believe that the team members on each process should have input on the problems they
face day to day because they understand them more than anyone. Team members at every level from the president down to
the line workers were asked to kaizen, which means to continuously improve their surroundings to make Toyota safer and
more efficient for everyone. This was all new to me at first,

but I began to learn the Toyota way and slowly got involved in kaizen activities.

Months passed, and I got my chance at permanent employment. I cheerfully accepted the reward for my hard work and was placed in my permanent position. Toyota's team environment made it easy to adjust and get to know my new coworkers. My team leader was heavily involved in kaizen and offered me the chance to get involved also. After production shift was over, he showed me some of the projects he was involved in. He had talent as a metal fabricator and told me he learned his skills over the past eight years from other team members and classes he had taken right here at Toyota. I knew how to use some of the fabricating machines like the drill press and band saw, and knew a little bit about welding. My group leader signed me up for class, and I became approved to use the shop equipment. I helped with various small projects over the next several months and used the experience to sharpen my creative-thinking skills. We improved the existing processes on line to make them even better.

My group leader approached me one night and asked me to try to provide a solution to a problem on line. I decided to give it a shot and accepted his request. He explained that a dolly that carried parts would not move with the line as it was designed to do and was creating an extra burden on team members because of all the extra walking on the process. As a line worker, I knew how much stress extra walking would cause over the course of the shift and reduced time to complete the necessary steps to ensure the quality of the cars. This was by far my biggest challenge yet. I began by consulting my fellow team members on both shifts, sharing my ideas and listening to their suggestions. I gathered my plans and started working. I decided to build a pneumatic arm to push the problematic dolly with the assembly line as it moved. This would eliminate walking back and forth for the team member and would reset itself after each vehicle on the line. I worked for several months to complete the project. I struggled a lot but received support from my team leader and group leader. I used scrap metal and good used parts from other equipment no longer used to save the company money. Plant safety was brought out to inspect my project to make sure it was safe to use. They complimented the design and gave me permission to use it on the assembly line. My group leader expressed his gratitude and even showed it to the assistant manager in my department; she complimented my work as well. I felt proud after these compliments, but the best compliment came from another work on line with me. He approached me before our shift began several days after my kaizen was implemented. He explained to me how it improved the process and reduced the burden for him. I felt good knowing that I helped my coworkers on my line.

Continuous improvement is just one reason Toyota is such a strong company. Toyota believes so much in its workforce that it opens up doors to all of its employees to explore their ideas and discover talents just like they did for me. Even though I have already learned a lot, I will continuously strive to do better than I did the previous day; and even if every project doesn't turn out as good as this one, I know I'll have the opportunity to keep trying and constantly improve myself and my workplace.

Mike Perkins: Dave Foster's Group Leader

The dolly that Dave built exceeded all of my expectations. In the past, the dolly would unsync from the car while the TM was trying to shoot the tail pipe to the center pipe. This created an issue because it would pull the data cable out of the gun at a cost of around $600. In a one-month period, we bought three of these cables. Also, the potential for a TM to get injured was very high. The TMs didn't realize the dolly had released, and the gun would pull back on them while they had their arms extended to shoot the pipe. This kaizen has since eliminated that along with the ergonomic issue that was created due to the TM having to pull or push on the dolly.

When I gave Dave the project, it was an attempt to get him more involved in his work. I could have fixed the dolly on my own, but this wouldn't have taught him anything. The project was meant to be a training exercise with minimal supervision to help him work on his problem-solving skills. Dave has surpassed all expectations that I had for him when I gave him the project. He has become a teacher to his fellow TMs. He has worked on many projects for many different groups now. While he is working on those projects, he seeks involvement from all the stakeholders. He has them work on the problem along with him, teaching them how to solve them on their own.

Recently during a major line speed change for us, Dave played a major role in setting up the processes for the entire group. Dave laid out a proposal for the entire group that worked better than anything that we had come up with so far. I would like to say that I was the cause for his success, but respect for people and continuous improvement are the two pillars that the Toyota production system is founded on. When given the opportunity, people like Dave realize abilities that they never knew they had, and this is the reason Toyota is such a strong company. The TMs drive to succeed because they feel they have ownership in it.

> David Foster has a very bright future at TMMK. I have seen him grow as a fabricator and problem solver over the last few years. He was promoted in 2010, and it is going to be exciting to watch him continue to move up in the company. —Tim

"It is ironic that many American companies do not have these studies, but yet Toyota because of their diverse environment allows us to do this."

Ron Kristoff: Team Leader, KGPC Trainer

Transition to Toyota

October 6, 1997, was my starting date at Toyota. I didn't realize it at the time but found out later that some of the companies I worked at in my career were suppliers to Toyota.

I worked at several steel mills in northern Indiana, with National Steel—Midwest Division being the last steel mill I worked at for twelve years. After visiting stamping's coil field at Toyota, I found National Steel's name on some of the coils. I also worked at Johnson Control's Foamech plant here in Georgetown, where I found out that they supply seats to Toyota. Eventually I applied at Toyota.

My career started in the assembly production engineering department for eight years. Some of my major achievements while being in this department were the TMMBC project in helping to start the Mexico plant. I trained two engineers and many group leaders. I had also worked many months on a big kaizen idea for the Assembly 1 Engine line in getting cutter bolts for the H-frame × control arms. They had many cross-thread problems from the ED paint and weld splatter in the floating nuts of the H-frame. These bolts took care of almost all of these potential cross-threads.

Next, I transitioned to the safety/KGPC department and had many grateful times when I was able to help team members in getting ergonomic issues addressed. The GPC training helps the new or transfer team members learn how to shoot fasteners, make wire harness connections, install grommets, hole plugs, and so on very quickly and saved three weeks on the learning curve in attending this class. Also we provide hybrid training (advanced and certified) for team members coming into contact with the hybrid vehicles in Plant 2 Assembly to keep them safe. I have helped out on potential injuries in quality circles, keeping team members from future injuries in having body weld staying with the proper standard on weld splatter on the door-opening areas of all of our vehicles. I have also enjoyed working on many assignments in safety, including safety audits, videotaping, and safety evaluations including JT/SJT and TEBA.

To close, I have to say one of my biggest enjoyments I have had comes from attending the Bible studies that I have attended at Toyota. I have worked for many companies, but Toyota is the only one that offers this. It is ironic that many American companies do not have these studies, but yet Toyota because of their diverse environment allows us to do this.

This is why I believe very firmly that Toyota has been blessed thus far, and is one of the greatest, most trusted, and respected companies in the world!

> Ron has a unique job. He is responsible for all the initial training to our team members as they come into the assembly shop. His dedication to his job has helped ensure the quality of many cars through his training. —Tim

Blair Perkins: Engineering Specialist, Assembly

I began my career at TMMK after graduating from Virginia Tech with a bachelor of science degree in mechanical engineering. Because of my work as a co-op student at TMMK for 1.5 years, I was offered a full-time position before I graduated. I happily accepted the offer to work for one of the premier manufacturing plants in the world!

Working at TMMK is a fascinating and stimulating experience for an engineer. The entire facility can be thought of as one massive machine covering 7.5 million square feet. That's 172.2 acres! This massive machine has the capability of producing two world-class vehicles every fifty-four seconds! An engineer at TMMK is surrounded by virtually every technology one can study. From miniature robotics to massive stamping mills, every square foot is covered with technological wonderment.

A recent project I led along with Mark Myers, also an assembly specialist, is a good example of what we do at TMMK. During the launch of the new Venza crossover vehicle, it was determined the current fuel tank installation machine (fuel tank lifter) would not be compatible with the new vehicle. After multiple failed attempts by contractors, I was requested by management to lead a team to design, fabricate, and install a new fuel tank lifter. This project was going to require the "A-Team"!

I affectionately call a diverse, highly skilled group of our team members the A-Team, named after the hit 1980s TV show's do-anything ex–Special Forces group. Our team, composed of maintenance and kaizen team members, can always be trusted to bring my equipment concepts to reality. Used many times in the past for other difficult projects, I knew they could complete the mission.

Members of the A-Team are:

Denver Napier: Maintenance team member, fabrication, programming, and electrical expert

Tim Rutledge: Kaizen group team member, electrical and programming expert

Lonnie Craig: Kaizen group team leader, fabrication and welding expert

Mike Chiles: Maintenance team member, machining and fabrication expert

Brad Sallee: Kaizen group team leader, fabrication and processing expert

Because of a very tight deadline and limited funding, there was no time for in-depth engineering designs in the traditional sense. I started work on the new fuel tank lifter concept after discussion with the team members that use the equipment. Our team members always provide the engineer with the solution. The engineer only has to listen carefully and then design the solution.

After the system requirements were determined for the equipment, I produced concept drawings. These hand-drawn plans and sketches depicted the equipment and its operation in detail. Yet the drawings allowed for adaptation and interpretation of the design by the A-Team, a key component to the project's success.

With drawings complete, the A-Team and I met to discuss the equipment concept and hammer out details. By using teamwork and allowing everyone to voice their concerns and opinions, a final design was completed that everyone was in support of and agreed upon.

In a matter of a few weeks, the fuel tank lifter began to take shape. The A-Team split the work up, each team member using their individual skills, to achieve a common goal. When needed, we would discuss a design and redo components if needed. Our team always had the focus to give the assembly team members (our customers) a quality, professional piece of equipment that exceeded their expectations (a key Toyota philosophy). Because of the A-Team's high skill, craftsmanship, and unwavering pride in their work, we were able to achieve our goal.

With the new fuel tank lifter complete, the A-Team then installed it on the assembly line. The full-circle process of taking a piece of equipment from concept to an installed, functioning, and successful machine is very important. By completing the equipment full circle, the team develops strong ownership and pride in their work. Furthermore, the team's skill increases every time they complete a job. They also are 100% familiar with the equipment when it inevitably must be repaired and/or modified. The team "owns" the equipment and is responsible for its successes and failures. Not an outside, expensive vendor that is maybe hundreds of miles away and lacks Toyota's strong philosophies in quality.

After installation, the new fuel tank lifter was a great success for assembly. Because of the lifter's design, it virtually eliminated any burden on the assembly team member's shoulders, a significant problem with the old equipment. The equipment also fully installs the fuel tank, only requiring the team member to tighten the bolts that secure the tank to the vehicle. By being "full auto" and operating in approximately twenty seconds per car, the equipment decreases the team member's "non-value-added work time," making the process very efficient. And lastly, it achieved the original goal of handling and installing three different-shaped fuel tanks automatically.

The fuel tank lifter is just one example of the many projects we do at TMMK. By combining our broad knowledge and focusing on a common goal, we can truly achieve great things. So "If you're in trouble and no one else can help, maybe you can hire the A-Team."

> Blair and the entire A-Team did a great job with this project. It is just another example of teamwork at a Toyota plant. —Tim

Chapter 6

Our Thinking Minds: Stories of Quality Circle Activity

Many companies hire consultants to come teach them lean manufacturing and then wonder why it didn't work at their facility. Sure, they can save large amounts of money by using our kanban systems and reducing inventories. In most cases, I would guess that the work culture doesn't change.

Quality circles are a key to our work culture. We come together as a team to fix issues or reduce cost. For example, last year I worked with a team that wanted to do a simple theme. The idea was to fix some small imperfections in the floor. We had previously hired a contractor to fix areas where concrete would chip away. These chips in the floor would grow and become a potential injury to a team member twisting his or her ankle. This group of team members began researching materials that were approved to use at our factory. Not only did they begin fixing the issues, but also they trained their fellow team members in other areas of the assembly plant to fix these areas as well.

This quality circle saved TMMK over $200,000 in one year. They improved their workplace and improved the safety of their fellow team members. The following are just a few stories of quality circles in action. There is over $1,000,000 worth of savings in this chapter, as well as stories of improving quality to our products to ensure our customers get the best product possible. Quality circles and problem-solving groups are just one system used to promote our work culture and are part of the magic of TMMK.

"You are constantly engaged in problem solving all day long without really realizing it."

Renee McIntosh: Team Member, Quality Circle Support

A Team Member's Day at Toyota

Being on second shift really gives you a new perspective on life. Working from 5:15 p.m. to 2:00 a.m., you have to change your whole lifestyle around. Working on second shift at Toyota has given me a new look at things I would have never been interested in learning and succeeding at.

In June 2002, I was hired at TMMK as a full-time team member, after being a temp for two years and seven months. A long time, I know, but worth the wait. Have you ever worked six days a week for $5.75 an hour and no insurance? After you get a job at Toyota, working second-shift hours isn't so bad because of the pay and benefits. If I have to work, I want to be paid well and have an incentive to do so.

Problem solving is a must and a lifestyle at Toyota no matter what shift you work. You are constantly engaged in problem solving all day long without really realizing it. Even in everyday life, we problem solve.

Toyota provides their team members with incentives for learning problem solving and applying it to our area. Incentives are money, competitions, prizes, and last but not least a smoother-running job that makes your night at work a little easier. We all want things to be easier.

So in 2002, I joined a quality circle on Chassis 3. A quality circle is a small group of team members who meet on a weekly basis to identify problems, investigate causes, implement counter-measures, and track and report results. A team member named Andrew approached me after a couple of months and asked me to join his quality circle, the Scrappy Do's. I was glad to join so I could learn what this place was all about. Throughout the next couple of months, we picked a scrap theme, gathered data, broke down the problem, and countermeasured the problem (theme).

It felt good to make a difference as a team member and not have to wait for group leaders and above to solve all the problems each process had. A place this size has many problems, and taking ownership and being a team player to help get them solved is really satisfying.

In November 2005, a team member named Marvin approached me and asked me to join his circle, the Fab Five. They wanted to work with a team member on Trim 3 to help Trim 3 solve the rear reinforcement issue on the Avalons. The parts were not fitting properly to the car. This process was in my team, and I had to do it at least once a day. The whole team hated this process! We had to hammer the reinforcement down to get it to fit to the frame of the body so we could install it to the car. Every other car was an Avalon. After two hours of this, you were worn out for the rest of the night. Truthfully, I think they got tired of hearing that hammer hit bare metal every other minute!

So we got started and tried to identify the problem. I never knew there were so many parts that made up a rear end of a car. Let's just say we learned a lot about cars on this theme!

We had to go to inspection, body weld, stamping, and the production engineering group to get info on all the parts. This process took us about a month to get through!

To make a long story short, we had to go all the way to the supplier to get this problem fixed. The issue was with a datum pin and a jig that were worn and causing the robot to weld the part to the car incorrectly, which led to the reinforcement not fitting to the car.

There was one happy team in Trim 3 for not having to hammer on the reinforcement every single night. Not to mention the team leader and group leader who had to repair the ones we couldn't get on!

All in all, I feel working at Toyota has not only improved my problem-solving skills, but it helps me problem solve outside of work in our day-to-day activities. I am thankful for TMMK and the programs that they offer each worker to engage in and learn and grow in the company, not to mention advancement. I have been here almost nine years, and I am still learning something new, every day. If you have the will and want, you can make things happen here. The Toyota Way is continuous improvement, as well as continuous learning everyday.

> Renee has a big job. She is the support team member for quality circles in Assembly 1. She supports nearly 600 team members with any obstacles they come across as they are making our plant better. The highest compliment I can and will pay anyone is this: someday, Renee, I hope to work for you. —Tim

"The triple As .. accept change, adapt to change, and accelerate over any obstacles."

Brian Staples: Team Leader

I have been working for Toyota for ten years and have been a functioning part of many sections such as chassis, quality gate, cart, and trim. I have always used quality circles to help solve problems with everyday activities to help eliminate or control unforeseen issues that can occur on a daily basis. In turn, this helps to maintain team member morale and their satisfaction of knowing they have ownership of their processes.

Quality circles are just another unique tool Toyota has given us to help solve problems dealing with indicators such as safety, quality, productivity, and cost. It also allows the team members the ability and knowledge to work together to solve problems dealing with their everyday jobs. Allowing the team members the ability to self-develop their problem-solving skills results in a sense of ownership of their processes along with the satisfaction of being part of the problem-solving process in a team environment. Being a team leader in Assembly 1 isn't always an easy task.

Working with the team that installs the luggage weather strips, team members were having difficulty installing the Avalon weather strips more so than the Camry weather strips. Team members were feeling discomfort, during and after running the luggage weather strip process, in their hands and shoulders. Working as a quality circle, each team member had an important role to play in problem solving the safety issues of installing the weather strips. Starting off with genchi genbutsu ("going to see"), we went to the process and began looking for variations between the Camry and Avalon weather strips. First, we started by taking push force readings in eight different spots of the trunk lid area. Our findings indicated that Avalon push force was far greater than the Camry. Then we looked at the thickness of the lip of the trunk area, finding out that it was in standard and both models were close to tolerance for the thickness.

Next we investigated the stiffness of the weather strips; we noticed that the Avalon weather strip was more rigid than the Camry. Checking the area of weather strips that set to the lip of the trunk area, we cut the weather strips apart. We then used a caliper to check the inside width of the channel that is pushed on the lip. The widths of the inner channels of the weather strips were very close to the same on both models. We then checked to see if the correct amount of mastic was put in each weather strip in case the Avalon was too dry, which would make it more difficult to install. Mastic is a sealant used in the channels of the weather strips, which secures them to the lip of the trunk helping to prevent the weather strips from coming loose and raising up, causing water leaks in the trunk area. Both models had similar amounts in them, meaning that the channel of the weather strips wasn't dry, causing the installation problem.

The weather strips are delivered and stored in a heater barn so the weather strips will warm up and make for easier installation. Our next step was checking the temperature setting on the heater barn to make sure the barn was reaching the correct temperature and wasn't set too high or low causing the issue. I used an infrared thermometer to gauge the internal temperature of the weather strips. We found out that the internal temperature was far lower on the Avalon than the Camry by at least 10 degrees. While investigating, we found the Avalon weather strips hung closer to the front of the heater barn than the Camry. The fans on process were cooling the weather strips off before (we were) pulling them from the heater barn and installing them. The parts further back in the heater barn had a higher internal temperature in the channels of the weather strips than the ones closer to the door opening.

Running a trial using parts further back in the heater barn, we found that the push force readings were closer to standard and the weather strips could be installed much easier than before. By making small stoppers to recess the weather strips in the heater barn by four inches, this made the temperature stay closer to the same on both models. With the weather strips being easier to install, it helped stop the discomfort everyone that worked on the process was feeling in their hands and shoulders while on and after running the process. With the weather strips being easier to install, the team members didn't have to move faster to stay caught up on the process, thus helping to reduce the physical burden of performing this process, therefore giving more time to check quality of the weather strips to ensure they were fully seated. This also helped to reduce quality issues with the weather strips not being fully seated, causing water leaks. Just by using the stoppers to recess the weather strips, we were able to keep from increasing the temperature of the heater barn. By not increasing the temperature, this helped to keep the energy consumption the same instead of using more electricity to heat the barn to a higher temperature.

Having everyone in the quality circle able to run the process, we utilized the knowledge they had to help solve the problem. Ultimately, we were able to reduce team member burden during installation while improving quality by reducing water leaks caused by out-of-standard luggage weather strips.

I believe the main reason that quality circles are so important is the way team members have the ability to (the triple As) accept change, adapt to change, and accelerate over any obstacles they come across on a daily basis. This is just one great example of how teamwork and dedication can help drive Toyota to continued success while "moving forward."

> Brian is a team leader in our trim section. His group installs the head liners, visors, and air bags along with many other parts. Brian takes his job seriously, and Toyota is better because of it. —Tim

"This is just an example of how quality circles work for all of us."

Mark Alexander: Team Leader, Body Weld

Quality circles give us a chance to improve the safety, quality, and production in our area. It also gives the team members a chance to get out and visit other areas and meet other team members that they may need for support.

We recently dealt with a problem where team members had to use a hammer to remove masking caps from the fuel tank. This was a huge safety issue as well as a quality issue. The impact from the hammer also damaged the caps. This cost the company money to buy new parts.

By forming a circle, we tackled the problem head on. The circle members learned how to use the problem-solving tools to work through the problem. We fabricated a tool that would help us remove the caps safely and without damaging the masking caps. We were also able to work with Maintenance Headquarters, to make replacement handles instead of trashing the caps. The circle also yokatened the tool to our other body weld areas. This is just an example of how quality circles work for all of us.

Greg Ballou: ESI and Safety Team Member

I was called to Mark's group concerning an injury to a team member. The team member had struck his thumb while trying to remove a sealer cap. The team member was not seriously hurt; however, the potential for injury was high. Mark's QC (quality circle) group had already been looking into the problem and had just placed in a lever to be used in the place of the hammer.

The group was using a hammer to strike the sealer caps placed on the fuel tanks. The cap is to keep paint from being sprayed onto the surface area of the fuel pump. A team member must remove the cap for another team member to install the pump unit.

The group had made a change in the way tanks were being sprayed due to production needs. In a 3–4 month span, they had replaced about 50–60 sealer caps at $105 a piece. Now they also had an acute injury due to the change. The sealer caps could no longer simply be pulled off, due to dry time. A check was done on the caps; a straight pull by hand would require forces between 75 foot lbs to over 135 foot lbs. The lever, installed and designed by Mark and the QC group he leads, required only 12 lbs.

The quick actions and leadership shown by Mark, who worked with team members and day shift counterparts to kaizen and countermeasure other problems, have minimized damage to sealer caps and fuel tanks, reduced cost, and eliminated the risk of injuries to the team members. The QC theme was also presented to NIOSH and the ERGO Cup. The theme or presentation will be in competition with others from companies such as Boeing, Delta, and Honda.

Toyota's belief in its diversity, respect for its people, and desire for continuous improvement have given people like Mark an opportunity to excel and use their abilities to help not only build Toyota into a successful manufacturing company, but also a world leader that other companies strive to match.

> Mark's kaizen has proven to be successful. Greg Ballou works in safety in body weld. He is very involved with his team members and their safety. Thank you both for participating. —Tim

"Your job is what you make it."

Steve Turley: Team Member, Quality Circle Support

Your job is what you make it. When working for other companies, this wasn't always true, whether it was due to poor management, lack of knowledge, little interest in the business, or just a plain bad attitude on my part. Right out of college, getting up and going to work was worse than studying for that calculus exam, dreading the moment I would walk through the door and start the workday. Eight hours would last an eternity, but the time in between shifts would speed by like a Kevin Dunn fastball.

Then I was given the opportunity to work at Toyota beginning in June 1998. Starting as a temporary, I came to work every day in Georgetown carrying that same attitude, and worked for around sixteen months before being hired fulltime. The first few years weren't so bad, focusing most of my thoughts on my personal life (newly married) while going through the motions every night at work.

One night, complaining to my team leader, it was finally drilled through my head that I could change the problems I was having at work. Toyota gives each and every team member the opportunity to improve where they work by using many different avenues, but mainly quality circles. A circle can pick the problem they want to work on, decide when they want to meet, and even choose how to fix certain items under their control. Instead of complaining about problems, I should take a little ownership in my job and try to improve it the best I can. The work wasn't the problem; the attitude was.

After eight-plus years of being in circles, I have learned an enormous amount about the auto industry and Toyota. More importantly, I have learned how to solve my own problems at work (with the help of other TMs) and developed numerous friendships from interacting with people throughout the assembly shop. Additionally, I have currently been given the opportunity to work in a "special projects" position in Assembly 2, with my sole purpose being as a resource for team members who are participating in quality circles themselves.

It is amazing to me some of the ingenious solutions that are created by team members throughout the plant. Not only does Toyota allow these people to operate, but they promote circle participation and strive to have everyone involved. On one hand, this is a very costly initiative that Toyota utilizes to stay ahead of the rest of the industry (overtime pay, theme completion, compensation, etc.) totaling hundreds of thousands of dollars. But when looking at the proven results of the QC program, mainly team member development and continuous improvement, it is just one more reason why Toyota remains in the lead of the world auto market.

> Steve is another of my past team members that I am proud of. He has a job now that fits his skills perfectly, supporting others to achieve their goals. —Tim

"Everyone who worked in our plant should walk out
the same doors the same way they walked in them."

David Eads: Team Leader

I would like to share a project I worked on in the beginning
of spring 2007. At that time, I was working in our assembly
conveyance group. I delivered parts to our line side groups
who assemble our vehicles. A group of team members decided
to work on some safety issues within our plant.

We decided that everyone who worked in our plant should
walk out the same doors the same way they walked in them.
Accidents should never happen at TMMK. Our safety department started a conveyance grassroots initiative to address and
develop safety ideas. This small group grew in number and
had support from team members up through managers in all
our shops. We began with twenty-two team members; within
weeks, the number grew to over eighty.

Our grassroots separated into teams we call "quality circles."
We all worked with a common goal in mind to improve safety
at TMMK. My circle focused on training for the entire plant.
We already had safety guidelines for pedestrians and powered equipment operators. This project would expand and
better define our guidelines. It would make them simpler to
understand what was expected by pedestrians and equipment
operators in any given situation. We made our walkways more
visible and clearer. We added guardrails and gates to keep
pedestrians safer. Body weld had recently done this with great
success, and we wanted to use the same activity in our shop.

In one aisle alone, over 6,000 visitors and team members
travel it every day. At the same time, nearly 700 turns across
this pedestrian aisle are made by powered industrial vehicles.
It should be noted in our plant that vehicles have the right of
way. Accountability is held by our drivers to make safe decisions. We produced training literature to inform the team

members on our safety policies and their responsibility to follow them. We also worked on a project to change our aisle way colors. Each shop at that time was different. We wanted all the shops to be standardized.

Another circle worked on having individuals wear high-visibility clothing that is required to work around powered equipment. High-visibility boundaries were set around our dock areas. This was done to keep visitors and team members aware that extra precaution is needed in these sections. This project has been accepted as our standard at TMMK, and hopefully all of our North American plants will soon follow.

We needed welders to build guardrails, so safety had forty-plus team members go through a welding class provided by the company. This training has allowed us to branch out and fix many issues. Many other quality circles were started and implemented across the assembly plant. At one time, we had over 200 team members working on safety concerns. We ended up saving our company millions of dollars in projects that reduced injuries, walk time, improved quality, and so on. We have greatly improved our safety in our plant. I am proud to say I was a part of something so important.

> I had the opportunity to work with Dave on this project. This is when I became a very big fan of his. Dave has now moved on to work in a special project group. —Tim

Chapter 7

Our Family: Making It Personal

Every factory has many families that work together at that facility. TMMK is no different than the other companies that move to a new area. TMMK changed our region. The excitement from the good pay and benefits caused many families to want to be a part of the success. My mother and brother both work at TMMK. My mom's story is in this chapter. At first, I was against her working in a factory. I believe every young man hates the idea of his mom doing this kind of work. I can tell you that I am proud of her. She works hard every day delivering boxes of parts to the lines. I have seen twenty-one-year-old men complain at doing this kind of work, but she continues every day to get the job done and always has a smile on her face. She is an example and inspiration to many, who look forward to seeing her driving her tugger down the line.

There is a saying "It's just business—don't take it personal." Or "Don't mix business with personal." I despise these sayings. We all have to make it personal. The people at the top have to look out for the people at the bottom, and in return the employees have got to have the best interests of the company in all our decisions at work. Any company or employee who doesn't follow these principles is doomed to fail.

Having families work together makes it personal. Building cars that we know our family members drive every day makes it personal. That is the key to success for anyone in any profession.

"My challenge to each and everyone at Toyota and around the world is to have an 'attitude of gratitude.' You too will see that you are truly blessed."

Lillie Turner: Team Leader

Blessed beyond measure: yes, that is what I am. The youngest of eight children; one child was stillborn. There were many challenges in life trying to keep up with all my siblings. I attended a one-room school for six years in Breathitt County, Kentucky. I was privileged to be a part of history when I presented a puppet show for the first lady (Lady Bird Johnson) when she came to our school to turn on the lights for the first time. The Johnson administration had put forth a lot of effort to bring simple everyday items like electricity to eastern Kentucky in the 1960s. I was once told that we were poor, but I never felt poor; in my heart, I have always been rich. I had the best of parents. I never went to bed hungry; I always had shoes to wear and a roof over my head.

My mother was my first teacher in diversity: she said, "Love everybody," and she lived it.

My dad was my hero; he was my protector and provider. He would work in northern Kentucky (nearly three hours away) and drive others to that area of the state. This was the only way he could provide for his large family and keep us in eastern Kentucky near the rest of the extended family. He taught me to live by the Golden Rule, be honest, have integrity, and have respect for others. He was a generous man that was always there to help when friends and neighbors needed help. He taught me to live with an "attitude of gratitude."

In 1973, I married my childhood sweetheart. My blessings continue to flow. In 1974, my oldest son was born; in 1978, my second son was born. They are my pride and joy.

In 1995, my oldest son went to work at TMMK the week
before his twenty-first birthday. He worked really hard to get
the job. It took him nearly three years to go through the hire
process. I was impressed at this place that was so particular
about whom they would hire to build these cars. In 1996, I
accepted the challenge to go through the process and get a
job there. At the time, I had been a dental assistant, so the
idea of working in a factory was a great challenge. I went
for the first test. I went home and told my husband, "I might
as well forget it. I am sure I didn't pass the test." I then got a
letter saying I had passed and that I would be contacted for
the next step. Every time I received something from TMMK,
I responded with a "Thank you" and "Yes, I am still interested in working there." In April 1997, I was at the hospital
in Hazard, Kentucky, with my mom who was giving up her
long battle to live. Her doctor had given up on her eleven
months earlier, and it had been a difficult year. The call to
work at Toyota could not have happened at a better time. My
hire date was April 21, 1997. Blessed, yes. Challenges, yes.
I told my second son, "You should put in your application."
He did; in November 1998, he started working at TMMK.
Toyota would become a family affair for us much like it is for
so many Kentuckians. The three of us have a total of thirty-
six years of Toyota experience. Working at TMMK is both
demanding and rewarding. It requires dedication to always
give 100%.

It is our job and goal to build the best cars in the world. I
have always said that the greatest asset that Toyota has is its
team members. I work with some of the best people in the
world. I care for their well-being; I encourage each one of
them to kaizen their processes to help them stay healthy.

I am thankful for Toyota and all of the people in
Georgetown who chose to build here in the heart of Kentucky.
TMMK provides a large group of people with a good income

and benefits that allow for nice houses, a good education, and many other things for their family.

> What can I say about my mother? She is the best; she is my inspiration; I am proud to be her son. I live each day to the best of my abilities because of the example that she and my dad have taught me. —Tim

Abbie (Age 9) and Cash Turner (Age 7): Children of Tim Turner

Our daddy works at Toyota. We get to do many fun things because he works there like Kings Island. Every year, we have Kings Island day where all the Toyota kids go to play together. Our favorite ride is the bumper cars.

When we were little, we had an Easter egg hunt at our house with our daddy's friends from his group and all their kids. Our deck was full of people and kids. We played on the swing set and searched for eggs. The kid who found the special egg got tickets for the Cincinnati Zoo.

Because of our daddy's job, we get to go on lots of fun vacations like to Disney World and get to go camping.

Every year, Toyota Safety has a coloring contest where we get to color pictures. The kid with the winning picture gets to be president for a day at Toyota. It is a fun contest we do during National Safety Month in June.

Thank you, Toyota.

"So now, there are two generations of us working here."

Bob Ditty: Team Leader

I can remember back in 1989, I kept hearing about the Toyota plant in Georgetown. I remember one day, I looked over at the guy working next to me in the coal mines. I told him how I would love to be able to work at the Toyota plant someday. Well, finally, after a lot of tests and praying, nine years later, it came. I got the call that I had been waiting for.

I had various jobs in the mining industry, surveying, carpentry, and drafting. These jobs require being done in such a way that there was no room for errors. This also applies to building a great car: no room for errors. I depend on all the team members to feel the same way. I've purchased one of these great cars for my wife, Vivian. SHE LOVES IT!!!

Working for Toyota has provided me with a great opportunity to work with a lot of gifted people. Our quality circle was solving a problem on our line. We received a great deal of help from another quality circle to achieve our goal. This showed me what can be accomplished when you get a team to come together as one for a common purpose.

Toyota has also provided my family and me with a great way of life. As parents, we always want the very best for our children. To watch them grow and become adults we can be proud of. I have always been so very proud of my son, Brandon. He has grown into a very good man. The one thing I always wanted for him was to be able to get a good job. To be able to support his family and have the things in his life that he desires.

Working at Toyota has not only allowed me to provide but also has allowed my son to do the same. Brandon was hired as a full-time team member in 2007. He works in Body Weld 2. So now, there are two generations of us working here. I cannot find the words to describe how I feel to know that he has this job.

To be able to say that my son and I both work at TMMK— it's GREAT!!!

Someday I hope and pray that one or both of my grandsons and/or my granddaughter will be able to work here as well. That will make it three generations.

My son and I are both so blessed and very thankful for our careers at TMMK.

> Bob Ditty passed away on September 5, 2009. He was a great person and friend. As you could tell, he loved his family and coworkers. Anytime we lose a friend and coworker, it creates a void at TMMK. Bob is no exception. A day doesn't go by that I don't think of his friendly smile and compassionate nature. He always had time to help a friend. He will be missed. —Tim

"Thanks to the dedication of all the team members, from engineering to the line workers building the cars. This car protected Austin's life. For that, our family shall be forever grateful."

Zenith Sorrell: Team Leader, and Jeff Sorrell: Skilled Team Leader

Many times in our everyday lives, we take for granted how important our jobs at Toyota are and the impact they can have on our customers' lives. This came close to home on March 21, 2008.

At approximately 1:00 a.m. on March 21, I received a late-night phone call. This is the call we all dread: something had happened. My sister was on the other end of the line hysterical, informing me that my nephew had been involved in an accident. On the way to the hospital, I came upon the crash site and immediately lost all hope.

When I stopped and got out of the car, my knees went weak. The car was unrecognizable. I thought if he wasn't dead, he was badly injured.

According to local law enforcement, who recreated the accident, Austin's 1998 Camry had dropped off the right shoulder of the road, he had overcorrected, and the car flipped three times sideways, and twice end to end before coming to rest on its top. The luggage area of the vehicle was now compacted up under the car. The front end had disintegrated. A, B, and C pillars on both sides were still intact, protecting the cabin area. It was unbelievable.

When I arrived at the hospital, I expected the worse; miraculously, Austin had only minor cuts and bruises, despite not wearing a seat belt.

The crash has given Jeff and I time to reflect on our jobs and the vast impact we as team members have. Every car we manufacture has so much effort put forth to ensure the highest quality and safety standards possible. From the initial design, engineering, suppliers, and manufacturing teams working together, we have a product that we can all be proud of. This gives the team members an opportunity to put a name, a face, and a story—this time with a happy ending—to their job every day, that what they do does matter.

The tan 1998 Camry was produced by TMMK Assembly 1 on August 6, 1997. It lined off on first shift at 3:42 p.m.

Thanks to the dedication of all the team members, from engineering to the line workers building the cars. This car protected Austin's life. For that, our family shall be forever grateful.

> Jeff works in Assembly 2 maintenance, and Zee is a team leader on Trim 3 in our plant. I am very glad that your nephew was safe. Thank you for sharing this story. —Tim

"One thing my daddy used to tell me is 'Little girl, if you smile the world will smile with you, but if you frown it brings everyone down.'"

Kim Burnette: Team Member

I have been married for twenty-three years, and we have four children. I work in Assembly 2 Final 2. I have been on this line almost two and a half years. My clock number is 18176. It took me two years and eight months to get that number. I was a temporary for almost three years.

In this time, I was fortunate enough to work in Body Weld 1, Assembly 1, and Assembly 2. Fortunately, my husband worked in Body Weld 1 and has been there almost fifteen years. While I was in body weld, we were able to meet each day for lunch, and we only worked one line over from each other. Mark, my husband, works in Body Weld II pilot; he is a team leader now on day shift. We used to live in Louisville, but when he was promoted to team leader and had to go back to second shift, we decided we were going to sell the house and move to Georgetown. We moved here to Georgetown in 1999, and my mother-in-law moved here with us.

Our family is the MOST important thing to us. So when I was hired and had to go to second shift, we both wondered how this was going to work. My husband was also on second shift at the time. It worked out good, Mom living with us. Someone we could trust with our kids. She lived with us in our house until August 2007, when she passed away from lung cancer.

From the first day we found out of Mom's diagnosis, the family of Toyota was very supportive. Letting us have time off when we needed it and just being supportive. When Mom passed away, group leaders from both our groups came to the funeral home. Several of our friends and coworkers from our groups came also and stayed with us as we went through this awful time. Some just popped in for a minute before work to say they were thinking of us. But we appreciated everyone who came.

One special lady in my group even brought my whole family dinner. She made the whole dinner, including dessert. Everyone was so nice and thoughtful. Toyota sent a beautiful flower arrangement to the funeral home, and sent several platters of food for the family. They even came to the funeral home to make sure we had everything we needed. Toyota is a special place to work because that is my second family. I talk to them every day about good and bad times. We all have nicknames for each other. They know my moods, and I know theirs.

I have had several jobs before I started working for Toyota. And I always feel that people are what make the job. You could have the best job in the plant, but if you don't like anyone you work with, it makes each day harder and harder to come into work. I am very fortunate to be in Assembly 2 Final 2. In my opinion, these are the best people I have ever worked with.

One thing my daddy use to tell me is "Little girl, if you smile the world will smile with you, but if you frown it brings everyone down." I try to live by this rule and try to leave my troubles at the door. But it is not always easy to do that, so I sometimes turn to my Toyota family to get me laughing again, which they always do.

Two of our children even worked here for the summer program. My son, Jeremy, worked in body weld, close to his dad. And my daughter, Heather, worked in Paint 2. They both enjoyed the experience. It also showed them what Mom and Dad do every day. It is important for kids to know the value of a dollar, and at Toyota they worked hard for their money. They were headstrong after leaving our plant to stay in college. If I could say one thing about Toyota, it would be "We have been put here on this earth for a short time, and we never know when our time will be up. Having a good home life that is able to be cared for, because of our jobs at Toyota, and having a good work life, with good friends and respectful leadership, also because of Toyota, makes our time here blessed."

May God bless you and all your family members.

> I had the opportunity to work with Kim while in the final section. One of her jobs is to install the door trim. She is one of the most outgoing, caring team members that I ever had the pleasure of leading. She had the opportunity to be in a Toyota commercial. I promise she has the right personality for it. —Tim

> "I became a father and began to think about my
> daughter's future. I wanted her to have as many oppor-
> tunities in education as possible to better her life."

John Booth: Team Member, Quality Gate

Why I work at Toyota...

At the early age of thirteen, I began my first part-time
job in eastern Kentucky at a local gas station that my uncle
owned. I pumped gas for customers and performed various
odd jobs. Five years later, a coal company opened a mine
in my community, giving me the opportunity to make more
money as a coal miner. I left the gas station and started out
as an underground equipment operator, where I ran a shuttle
car and continuous miner.

Many are unaware of how dangerous coal mining can be.
Not only did we risk the dangers of mine cave-ins, equipment
malfunction, and explosions, but we also faced the nearly
inevitable lifelong side effects of black lung disease. Living in a
rural area without a college education, coal mining seemed to
be the best way to earn a decent living.

As years passed by, I began to realize that in order to
progress in mining, I needed to take classes to become a section foreman. My duties would then consist of leading twelve
fellow miners in daily operations. This job offered me opportunities in management and also included a larger salary. In
addition to this, I began taking college courses to become an
EMT. That helped me become a part of the mine rescue team.
This group was involved in recovering people in mine trag-
edies nationwide.

I became a father and began to think about my daughter's
future. I wanted her to have as many opportunities in education as possible to better her life. My daughter is currently a
freshman at the University of Kentucky, where she is working

to obtain her bachelor's degree to become a physician's assistant and is currently on the dean's list.

Toyota gave me this chance. I started working there as a temporary in 1999, which led to me being hired as a full-time assembly worker. Now I work in a quality gate group. It is our job to inspect the cars as they are being built. We are there to ensure that quality and safety are built into every car before other parts cover up the very important components that are underneath all the garnishes and trim. It is our job to ensure that our customer gets the most reliable, safe car built today.

> John Booth is a dedicated father, and he brings that dedication to work with him. It is John's job to ensure that quality is being built into every car, and he takes his job seriously. —Tim

"The day care has given my son the foundation he needs to be successful when he starts school this fall."

Deanna Feeback: Team Member

My son was six months old when I decided to apply as a temporary at TMMK. Little did I know that starting a career with this auto-manufacturing facility would be one of the best things I had done in my life thus far. I began working in the paint shop as a temporary on first shift in paint finish. After eighteen months of first shift, I was moved to second-shift Assembly Plant 1 on trim line. It was a shock to my body: working at night, having my little boy at day care, and having my husband on second shift proved to be a real test to our marriage. Needless to say, the marriage fell apart, but I was left with something that means the world to me: confidence.

For the first time in my life, I was confident that I could make something of myself by myself. In November 2006, I was hired as a full-time team member. I had my own health benefits, a salary that I could use to start my life over, and confidence that I could provide a good life for my son and me.

Toyota has an on-site day care that I love. My son wakes up in the morning and can't wait to go "see his friends." I feel very blessed that this is an option for working parents at TMMK. I can go to work knowing that my son is five minutes away and in very good hands. The day care has given my son the foundation he needs to be successful when he starts school this fall. I am very impressed with the knowledge he has gained throughout the three years he has attended the day care.

I am very proud to be employed by TMMK. I have great benefits and job security. Job security means a great deal to me, especially in today's struggling economy. I am very fortunate that I have a job, a stable income, and great medical and dental benefits. As we face the unknown in the auto industry,

I rest assured that Toyota will do what's in the best interest of the company and its team members.

> Deanna is a hardworking team member. She is going to do well because she always wants to learn. Thank you for sharing a story about our day care. It is an important part of Toyota's success. —Tim

Chapter 8

Our Education and Training: Reach for the Stars Program

I realize that a lot of companies offer tuition reimbursement. Our Reach for the Stars Program is dedicated to supporting each of us who wants to finish or begin school. The following stories are of team members who have taken that education and bettered themselves and the company with the degrees they have earned. I respect these people a great deal. To go to school and earn a degree while working in the fast-paced environment that we work in is amazing to me.

The really great thing about this is how they use what they have learned to once again kaizen and improve our company. So, while I personally am proud of each of these team members for their achievements, I am more proud of the education they provide their fellow team members every day when they talk about what they learned. It is the magic of the teachings that these team members share with their coworkers that makes it great.

Mark Sanborn wrote a great book called *You Don't Need a TITLE to Be a Leader.* In this book, he discussed the concept of feeding forward. When we constantly feed information and knowledge forward, it generates a level of excitement in learning and development. Mr. Sanborn is correct; Feeding forward is a key ingredient to success.

"My hope is that as I continually improve and meet further goals, I may contribute to Toyota's continuous improvement as well."

David Barnhart: Team Member

I value many things, but two things I value prominently: continual learning and continuous goal setting. Toyota enables me to practice both.

Prior to coming to work with the temporary agencies and eventually TMMK, I was in business. My partners and I decided to sell the business, I started work for a temporary agency, and Toyota hired me in January of 2005. One of the primary reasons I got out of my previous business is that I wanted to achieve some formal educational goals and I wanted to pursue other goals I had set for myself. The business I was in provided the means to attain some goals, but it was time to move on to others, and Toyota is very helpful in the process.

Toyota provided me with tuition assistance to finish a degree I started over 28 years ago. Because of circumstances at the time, I did not finish my degree. Coming from a family that values education and a family in which many members involve themselves in education, I had the impetus to work hard to achieve the degree even all these years later. My desire to learn and apply new knowledge spurred me further. Toyota helped make a way for me to finish my degree and realize one of my most valuable goals.

I learned formally about the management techniques and business practices of the company in some of my classes, but learning about them on the job helps me see the excellence and genius of The Toyota Way. Not only has Toyota provided a way for me to meet formal educational goals, but I receive an excellent business and operations education while working at Toyota.

I believe in continuous improvement. I believe it works for Toyota, and I believe it works for me and my family. I have many other goals and TMMK helps to provide the means to move toward their achievement. I am very grateful to Toyota for helping me achieve some of my educational goals and for providing a continuing education in my day-to-day employment. My hope is that as I continually improve and meet further goals, I may contribute to Toyota's continuous improvement as well.

David Barnhart impressed me from day one. He has a bright future. —Tim

"The reason I am able to complete my needed training is because TMMK encourages courses and offers them onsite through multiple programs including Reach for the Stars."

Keith Claunch: Team Leader

When asked where I work, I am proud to say Toyota Motors in Georgetown, Kentucky!

I believe that if you find someone with an internal drive, you should give that person what he or she needs to improve conditions around him or her. The advantage to this thought process is a lot of input from shareholders, quick changes, and recognition. All three of these factors make it easy for people with a good work ethic and the drive to take ownership of a project and improve not only their own area but also others that need help as well. These factors are also what I consider my drive to make TMMK a better place to work. I have been doing kaizen (constant improvements) at Toyota for the last eight years. TMMK has allowed me to become somewhat of an unofficial project leader for assembly.

I started out doing small things such as rebuilding flow racks and small fabrication repairs. Those projects quickly became larger and more complex. The reason for my project growth was good results from prior projects, and the willingness of TMMK and myself to acquire training in the areas I was weak. I came to Toyota knowing how to do some minor fabrication; now I am proficient in welding, pneumatics, hydraulics, and machining. I have also started my electrical, motor controls, and PLC training. The reason I am able to complete my needed training is because TMMK encourages courses and offers them onsite through multiple programs including Reach for the Stars. This led to one of my projects being entered in a national ergonomics competition. This is also a driving factor for me because of my competitive nature.

Several years ago, my team leader told me something that has stuck in my head: "Make yourself necessary." That comment has been one of my motivators since I started at TMMK. What started out as self-preservation has turned into trying to make TMMK the best and safest workplace it can be for all team members. There is no better feeling than knowing a team member was struggling on a process and I was allowed to make it better for them. TMMK is open to all ideas to improve productivity, safety, quality, and environmental issues. They encourage team members to bring attention to areas that could be improved and also be a part of the solution. If TMMK feels we have a valid concern and a good solution, they are MORE than willing to allow us to resolve our problem and give us the resources to do so.

> Keith is one of the best fabricators that I have ever had the pleasure of working with. He can design and build anything. His skills and knowledge have saved TMMK at least $1,000,000 over the last year. You are a true asset to the company. There is a patent pending on one piece of equipment that Keith has designed. —Tim

"They truly stand behind what they say they will do."

Orcelia "Lou" Spagnoletta: Assistant Manager, Toyota's Mississippi Plant

Looking back to my childhood, I realize people truly go through things in their life to make them stronger as adults. As a young girl, I knew I had to start getting my family from being so poor to at least middle class. For some reason, I knew I had it in me to change my life and my family's life. I wanted to have a much brighter future for myself, my family, and especially my son. It was up to me to get to where I needed to be without anything hindering me. I can remember as a young girl with three other sisters and a stepbrother that we did whatever it took to get food and clothes on our backs and food in our stomachs. My father worked it seemed like from daylight to dark trying to make ends meet, and my mother had to stay home and take care of all of us. I think at the time he might have brought in about 400 dollars a month, which isn't much considering how many children they had to feed and clothe. I remember my sisters and I would wait until the local landfill would close for the day, and we would sneak up there and rummage through bags of garbage just to see if maybe someone threw something good away that we could use. I also remember the Butternut bread truck bringing in the bread that was old and throwing it away and we would take it home and eat it. What a treat, we thought! Especially when we got some of the cakes and doughnuts.

So it is true to the statement another person's garbage is another person's treasure. There were so many times I remember thinking to myself, I don't want to live my life this way; I want more. When I started high school, I also got to play basketball for the Boyle County Rebels. I really think this is where I started realizing what it took to get out of the rut that my family was in. With the help of all my teammates and basketball coach, I owe them a huge thank you! Each and every person on our team and their families took me in and made me part of their family. The good thing about this is when I went to some of their parents' home, I thought, "Oh my God, how nice is this?" I told myself this is what I want when I get older. I would spend many holidays at one of my closest friend's house at Thanksgiving and other holidays (of course, that was after we would have something at my mother's house).

After seeing what everyone else had, I wanted to make sure some people did not think I was as poor as I really was. My brother and I would go work in hay, tobacco, and I even wrung a chicken's neck just to make extra money. I would save it all up and go out and buy myself a new pair of Nikes. At the time, I really thought this is what people looked at was your outer appearance. Little did I know they saw something more in me than that, and this is why my basketball coach, Mr. Stewart, surprised me one day with my letter jacket.

He and our statistician knew that I would probably never get a letter jacket if they didn't get it for me. It was one of the happiest days of my life. I remember crying and thanking God that we have people that have such great care for someone other than themselves. I still have that jacket. So all my life I have strived to make my life richer in every way. As I grew older, I knew things like this don't come to someone; you have to go to it to achieve it.

I remember talking with my mother, and she told me that she was so proud of me being the first one in our family to go to college. She said you need to do that or get a job that pays really well. I knew what she was telling me and knew what I needed to do. After high school I went to college and worked. I had to do whatever it took to make my way because I didn't have anyone to help me now that I was on my own.

My career at Toyota started in 1995, and it's one day I will never forget. I had put in my application about three years prior to that and finally got a call one night when I was working at UPS as a part-time supervisor. I had a personal phone call that came to me and I took it, not expecting to hear what the other person on the other end had to say. We would like to offer you a position at Toyota if you are still interested. My heart started pounding and I said, "Of course yes, yes!!!" I remember going up and down the feeder lines hollering and telling everyone that I got a job at Toyota. I was going to be a part of one of the best companies in North America. Was this my big opportunity to make my life different, or was this just another stepping stone?

Once I started working for TMMK, I placed getting my education on the back burner. My mom said I needed to do one or the other, and far as I was concerned I hit the mother lode. I knew that this job was going to take me out of the rut that all my family was in. I believe that God worked through me to not only help myself but to help my family out also. I was able to get a house to let my sister and her two girls live in and help her buy a car. I was also known as the favorite aunt because they knew at Christmas and birthdays they were going to get something really cool. It made me so happy to see my niece and nephews open their presents and see that huge smile on their face, because I knew how that felt when I got my "letter jacket."

As I started my career at TMMK, I was on top of the world for the first time. I knew that I would have some sound stability in my life. My first few months of working at Toyota, I knew working on line as a team member was not for me. The first chance that I could, I took team leader classes because lucky for us they promote from inside as well as outside. I had to come up with something so I could get noticed on the engine line. At this time was when process diagnostics was new for us and Wil James set a goal for us to achieve, and that was 90% efficiency.

I talked with my group leader at the time and told him that I wanted to be the first to accomplish this and use it as my presentation platform. I remember Wil coming over and I was nervous as a cat trying to explain what all I did to get to 90%. I guess I was successful because soon after that, I got promoted to team leader and was put on door line. After getting promoted to TL, I knew I wanted to keep climbing the ladder. I was then promoted to group leader in 2002. I started thinking about trying to finish my degree, so I inquired about the Stars Program and went back to school.

I knew that if I wanted to do anything more at Toyota and maybe pursue anything else at Toyota, I needed to finish. So I was working 12 or more hours a day and going to school. I was determined to complete my degree in business management. The last year I found out that my husband and I were going to have a baby. I had to take a year off, and I had a healthy baby boy. His name is Zachary! Wow, what a child does in changing your thought process. I was really determined then to make sure that I finished my degree because how could I push the issue about college if I didn't finish myself. I wanted to be a good role model to my son. Make sure, no matter what, continuing your education is very important. I am proud to say that I completed my degree in August 2009 with a bachelor's in business management. I could not have done this if it hadn't been for the sacrifices that my husband, sister, and mother did by rearranging their schedules to help take care of Zachary so I could go to class. Of course, Toyota having the tuition reimbursement program made it a lot easier financially to get it accomplished as well. This is just another reason why Toyota is one of the leading companies in North America, and how they want to help their team members better their personal lives. They truly stand behind what they say they will do.

So to sum my life and my Toyota life, I know we make a perfect match. Just as I was growing up and was inspired by many of my friends and family to reach my goals, so does the company I work for and that is to be the best company in the world.

In October 2010 Lou was promoted to assistant manager and offered a position at our new plant in Mississippi. She will do a great job and I wish her the very best. —Tim

"I don't know what is coming next with the way things are in the world. I believe that Toyota will be ready to lead the way in the automotive industry."

Arthur Black: Team Leader, Various Experiences

I started working for TMMK on February 4, 1991. I was hired into the paint shop and was assigned to anti-chip undercoating. Being new and not knowing a lot made the job a little stressful. I had some good team leaders, and they helped me learn my jobs and the different knacks that made the processes flow smoother. Other than working for my father, this was the first job that had mandatory overtime.

That was something else I had to learn to work through. I worked end-of-shift activities, which included paint line purging and paint/PVC filter changing. We put a lot of overtime maintaining the paint systems that we used.

Toyota management always encouraged team members to kaizen their processes to make them run safer and more efficient. I sometimes did this on my own by writing an idea up on a suggestion form. I was able to make some extra money from these suggestions. Other ways I was able to better the workplace was through quality circles. I helped complete themes that reduced waste and improved quality.

I moved from the paint shop to production control, where I learned about parts ordering and parts delivery. I learned to use some of the computer systems to track down part shortages for body weld shop, plastics shop, and assembly. Being able to find short parts fast is vital in the just-in-time system Toyota runs on.

In 1996 I was promoted to team leader and moved to service parts. There I learned how to ship outbound freight. I set up orders and loaded freight on trailers so it wouldn't be damaged. This time I started shipping parts to TMMC. Quality was very important. I moved from service parts to the north dock and got involved in parts ordering for the North American suppliers for Assembly 1. After a couple years at that, I was awarded a special projects job, which was parts flow and inventory recovery.

Four team leaders and the parts-ordering specialist maintained dock and line side parts flow. After that job was finished I went to Assembly 1, where I help maintain south minomi routes.

I have learned a lot in all the different jobs I have performed. They haven't all been fun or easy, but they have all been challenging. I don't know what is coming next with the way things are in the world. I believe that Toyota will be ready to lead the way in the automotive industry.

In 2006 I received my bachelor's degree thanks to the Reach for the Stars Program.

> Arthur has worked in a lot of areas throughout our plant. He is a good team leader. I have been told by his team members that they appreciate his willingness to support them. —Tim

"This group was set up with all of the Toyota KPIs in mind and was just like a production group in assembly."

Alfred Johnson: Group Leader

My name is Alfred Johnson, group leader of MA830. I have been a group leader for almost thirteen years and recently completed my degree in 2007. The degree I received is a BA in organizational management from Midway College. I participated in the Reach for the Stars Program in completing my degree, and I feel it has been very beneficial to my career here at Toyota.

I've worked in different sections and have had different groups. In the fall of 2008 I had the opportunity to help develop and establish a supplier supervised group in Assembly 1. This group is Employment Plus, which only had accounts for administration-type positions. This was a first for Employment Plus to hold positions on the floor in manufacturing. This group was set up with all of the Toyota KPIs in mind and was just like a production group in assembly. The group consisted of one group leader, one team leader and six team members on each shift. They are looked at as a mini-Toyota inside TMMK.

With the help of my college education, work experience, and the opportunities from my management team, Employment Plus was a great success, since previous attempts of establishing a contractor in the conveyance area.

Bonnie Smith: Assistant Manager, Conveyance

This challenge faced a lot of obstacles. Due to the economy, a stronger UNSEEN challenge with this group was the fact these team members were temps that were going to be hired by Toyota. It was, of course, a deep disappointment to them to be so close and then be cut off for a reason out of their control. Alfred was a great morale support to them at this time. Alfred put together a great plan to develop this area into a mini-Toyota. Truly his skills showed through with his plan. A bigger skill that is hard to teach was in his ability to encourage and support these team members in facing the reality of the economy. He challenged them to not give up and to support them in building up their morale and their knowledge in Toyota TPS. The integrity of these team members truly shined, and their deep commitment to Toyota along with the guidance of Alfred's skills allowed them to more than meet their expectations. During this project, Alfred was given the opportunity to use his college degree and got the satisfaction of supporting team members to rise above issues caused due to the economy.

> Alfred is one of the most humble people that I have ever met. He is an excellent group leader with a heart of gold. —Tim

Chapter 9

Our Teamwork

Teamwork—this book has it written all over it and through it. In most instances, the only way something gets done at Toyota is by applying teamwork. Sure, there are many competent people that are capable of doing a job and making decisions. The company doesn't see the talent in the individual—only the team as a whole. Our success or failures are based on the team concept.

Another aspect of teamwork is creating a welcoming environment where everyone feels that they are playing an important role in the success of the company.

"On my birthday, I was surprised to get a cake from a great lady who never forgets a birthday."

Andy Baysinger: Team Member

> Toyota's greatest asset is its team members.
>
> **—Eiji Toyota**

There I was twenty-four years old, with no family around, my wife pregnant with our second child, and I had just lost my job and consequently our house, which was owned by my employer. So I took a job with Precision Staffing at Toyota with no intention of long-term employment. I was desperate, and they were the first people who would hire me.

Taking a job at a car-manufacturing plant scared my mom to death ... she had heard stories from my uncles, who both worked for a car manufacturer, about how rough of a place they can be ... and honestly I was a little nervous myself. What would the people be like? What I found flew in the face of the stereotype.

My name is Andy Baysinger, clock number 45378, but who I am is a member of a team.

My first day on the job through a random lottery, I was assigned to Assembly, Final 2 (MA620). What I found there was not just a hardworking line but a community. I was expecting weird stares and some animosity because I was a temporary. What I found was a community woven together by hard work and shared experience instantly willing to embrace the new guy.

The moment I met my first team members they offered help....

"Do you know where to park? Where the café is? Let me help you find a locker. Your hands are going to hurt at first ... ice them down ... stick with it!" And I have. When my son was born, I remember opening a card from a team member and finding $100, which as a temporary was very much needed and appreciated. On my birthday, I was surprised to get a cake from a great lady who never forgets a birthday. When my truck broke down, team members drove out of their way to help fix it. Three years later the kindness continues; just last week, I left the lights on in my car and six people came out to give me a jump!

The French philosopher Tocqueville once said, "America is great because she is good. If America ceases to be good, America will cease to be great." What makes Toyota great are not just Toyota production systems, or Toyota business practices, or even the Toyota Way; what makes it great is that wherever you go … whatever shop you enter, you will find good people just like I found on Final 2.

As with any job, conflicts arise, but at Toyota an undeniable brotherhood remains a stronghold to our success.

> Andy is a humble person. Even though he won't like me saying this. He is also one of the most brilliant people I have ever known. —Tim

"In my study, they taught me the importance of keeping my eye on the goal and to keep on working toward my goal."

Jessie Childers: Team Member, Maintenance

Lessons

Knowledge is not wisdom, unless used wisely.

—J. D. Anderson

I was blessed to be the son of a coal miner from the mountains of eastern Kentucky. I grew up like most of the people that I knew from back home at that time, and that is "dirt poor." I never thought a lot about not having much. But my upbringing did teach me to be thankful for everything I am blessed with.

So I learned to be thankful!

After I graduated high school, I joined the United States Marine Corps. It was a rude awakening because I was ill equipped for the demands of military life. But, through much work and dedication, I became a Marine. I worked very hard, but no harder than the rest of my team members in my platoon. We learned the importance of working together for a common goal.

So I learned the importance of teamwork!

While still in the Marine Corps, I had the unique opportunity to train with the Japanese National Self Defense Force. I studied "Nippon Kakutodo" (Japanese Fighting Spirit).

It was the marital art used by the Japanese Army. While training, we were invited to fight in the All Japan Invitational in Mt. Fuji and then to Tokyo. In my study, they taught me the importance of keeping my eye on the goal and to keep on working toward my goal.

So I learned the importance of focus and persistence!

After getting out of the Marine Corps, I got a job with a major food manufacturer. While working there, I learned a trade. And after working there for a number of years, I had progressed to the height of my job. But in doing so, I became frustrated and dismayed because there was a set way of doing things and you could never change—not even for the better.

While not a positive experience, I still learned a valuable lesson. Don't be afraid of change. As a matter of fact, keep your eyes open for something to change to make things better.

And that brings me to Toyota. In April of 2000, my family and I were blessed when I got the call that I had a job here—to which I am eternally thankful. It has really changed my life inside and outside of the plant. Not only does Toyota embrace positive values in its team members, it encourages us to use them for the betterment of our lives and our business.

So, I have been able to take the things I learned in life: being thankful, the importance of teamwork, focus, persistence, and being open to change to the point of implementing it to make this a better place for us as workers and for the business as a whole.

My story is a little unique, but no more than all the rest of the stories contained here in this book. While we all have different backgrounds, I think the lessons we all learned in life have together made this truly a great place to work!

I am proud to be part of the TOYOTA TEAM!!!

On behalf of my family, I would like to thank Toyota for my job and the opportunity to go as far in life as I am willing to strive.

And, oh yes, one other thing I learned in the tournament fights: If the guy you are fighting has had a black belt long enough for it to be faded from black back to white, never, never let your guard down!

> Jessie is a maintenance team member in Assembly 2. One night he worked during his lunch break to help improve the safety of a dolly. I have since found out that he does this all the time. The dedication to your work is second to none. —Tim

"My dad once said, 'If you were to be a ditch digger, be the best ditch digger there is.'"

David Garvin: Team Member

Someone once asked me, "What makes you tick?" The first thing I thought of was God. That goes without saying, so I will try to answer the question in a little different way. There are several words that come to mind: family, pride, determination, quality, customer first, and tenacity.

Support from my family, especially my wife Debra, for taking care of things at home while I am at work. Her support allows me to concentrate on my work and quality circle activities. My oldest daughter Melissa is married to a wonderful man named Matt; she is a registered nurse. They are both very driven. I couldn't be more proud of the both of them and their accomplishments. My younger daughter Danielle is also married to a wonderful man named Kevin. They have blessed my wife and I with a very awesome grandson named Logan. Danielle is currently a stay-at-home mother, Kevin is in the Army. I am also proud of their accomplishments as well.

I learned at a very young age from my parents the definition of determination. To never give up and never quit. That if you are going to do something, whatever it may be, to always give your best. My dad once said, "If you were to be a ditch digger, be the best ditch digger there is." Also, way back in high school, my algebra teacher signed my yearbook with the quote attached, "Tenacity." I asked him what that meant; he said that I would question everything and not give up without an answer and that I would argue over a grade. Sometimes I would win, but like he would say, "I knew you would question me."

When I worked at UPS many years ago, I was a package sorter. We were taught that the next person down the line to receive the package we sorted was considered our customer. It was our job to ensure that every package gets to the customer in good condition, thus providing a quality product to the customer.

So these words—family, pride, determination, quality, customer first, and tenacity—are very much a part of what makes me tick at Toyota as well. I feel like the folks I work with are my extended family. We share stories and I have learned many things from them, as well; we support each other. I take great pride in knowing that I try to provide the customer with the best quality that I can, keeping in mind that not only are the customers who buy our vehicles but also the team members downstream are my customers. It is my job to provide them with a quality product so that they in turn can more easily provide their customer with a quality product, and on it goes.

Last but not least, quality circles. I am very passionate about quality circles. I have been involved since 2000. I feel that they provide a quality service to our customers. The circles also build teamwork as well as building problem-solving skills. It is awesome to know that you have the power to make life easier for our fellow team members by improving their work environment. It is also extremely satisfying to see the results from a theme, however small it may seem, because anything you can do to make your fellow team members' life a little easier will greatly improve quality of the product.

The one thing I try to pass on to new members is to be persistent and to have patience. My quality circle, the Checkered Flags, has adopted this motto: "We are there from start to finish." I hope I have answered the question, "What makes me tick?"

> Dave Garvin works on the engine line. He installs parts as the engine goes down the line before it gets installed into the car. He is an exceptional quality circle leader. His quality circles have won multiple awards. and they have greatly improved our company. —Tim

"The hard part is learning what the others around you are doing and how it's going to affect you."

Mike Bensing: Team Member

I can remember the guy that trained me at my first real job after high school telling me that he wasn't just going to show me how to do my job, but also how to figure out why I was doing it and how it affected everyone down the line. He taught me how to think about everything I did and how to work through any problems that came up. By thinking about why things were happening and if I didn't know why, I should ask questions of people that did. This is something that has served me very well during my twelve and a half years at TMMK.

Especially, the last ten years that I have spent in conveyance. Conveyance has very different work than that of the assembly lines because the work isn't as strictly standardized. Every time you go out to make a delivery with the same route, it is going to be a little different and you need to be able to think on your own how to handle the situation. When you train on a process, you learn where the parts need to be delivered and how they need to be loaded, but it is impossible to teach new team members about all possible situations that will come up. It's easy to learn your job. The hard part is learning what the others around you are doing and how it's going to affect you.

I have been very fortunate in my time at TMMK, and have met a lot of very helpful people who I have learned a great deal from, just by doing what I was taught early on. That is, ask questions even if it wasn't part of my job duties. Observing the maintenance men fixing equipment and asking them many questions about what happened and why. Taking a few extra seconds and watching the parts that I delivered being installed instead of just dropping them off and leaving. Talking to the guys that make the dollies that we use and learning how they are built and why they are designed the way they are. I have had a lot of classes and received certification in many areas here at TMMK, from welding and fabrication to first aid and CPR to fall protection and use of the one-man aerial cranes, just to name a few.

Most of the classes I have listed are not necessary for me to do my job, but they make things a whole lot easier, and I was able to receive this training simply because I was shown by one of my first team leaders how to look up the available classes on the computer and have had very supportive management that have allowed me to take them. It still amazes me that even after twelve years, I still see things all the time happen that I have never seen before, but because of not only learning what I was supposed to do but by being encouraged to constantly be thinking about what I was doing and how to do or make it better I manage to get through every day with very little difficulty. It is simply a fact that the more knowledge you have, the easier things are going to be. I am very thankful to everyone that I work with that has ever answered a question or taken the time to show me how things work or just let me look over their shoulder when they were working. One great thing about Toyota is we are always learning.

Mike has worked with many team members on projects. While he is constantly learning, he is also constantly teaching. —Tim

"World-class team makes a world-class vehicle."

Mary Jane Wills: Assistant Manager, Assembly 2

My life with Toyota started in 1992. I had worked my way through college at Kroger's and upon graduation I received a BS degree in education. As time went along my sister got a job at TMMK. I heard her speak the Toyota lingo: J.I.T. standard work, kaizen, andon, and the one I thought was odd was genchi genbutsu. I found it all interesting and very different from what I was used to in Winchester, Kentucky. She even got to go to Japan and talked about the training she was getting, and I thought to myself that it sounded like an interesting place to work. I started the testing process and got hired. I will always be thankful to Charles Luttrell; he was the one that interviewed me and recommended my hiring. Thank you, Chuk: for giving me the chance to work at TMMK and giving my family the experiences of being in the Toyota family.

When I announced I was going to work at TMMK, many people outside of here would say, you won't like doing the same thing day in and day out. I know now, they must have never worked for Toyota, because we are always changing and improving. I have met many wonderful people here and many great trainers. I have worked in groups that were like a second family to me. I have worked with team members from Iran, Brazil, Japan, Germany, Mexico, Sweden, and Africa, just to mention a few, and many more countries. I think that I have worked with someone from almost all the fifty states as well. I have learned about other people, cultures, and what makes us unique, also how we all come together to build a world-class vehicle by being a world-class team.

When I started to work at TMMK, I started as a team member in trim in Assembly 1. I was moved to Assembly 2 and was promoted to team leader. I was safety leader and a quality circle leader. Always looking to learn new skills, I became the team leader for trim during the first takt time change at TMMK. I was able to get promoted to group leader and have worked in trim, conveyance, logistics, and the cart repair area. I am now an assistant manager in Assembly 2. I helped start the first business-partnering group here at TMMK, the Women's Leadership Exchange Network (WLEN). It allowed networking and mentoring with other leaders of the company. It supports TMMK success by developing, supporting, and training. It also encourages an environment that encourages inclusion, while recognizing and respecting diversity in the workplace.

With it all said and done, I think it is the people who are so great here at TMMK. It is not just Toyota. It is not just an automotive plant in Georgetown. It is a group of people who work together, get to know each others' families, spend time outside of here together. It is a bond that is strong throughout our Toyota family. As I said before, what makes us unique is how we all come together to build a world-class vehicle by being a world-class team.

> Mary Jane Wills is currently an assistant manager in Assembly 2. The diversity she describes in this story is true to our Toyota way. I want to thank her for participating and sharing her story of diversity and teamwork. —Tim

Chapter 10

Our Understanding: What We Have Learned Along the "Toyota Way"

It amazes me what people remember. Some people unfortunately only remember the bad stuff. You can do 100 things right and never hear a "thank you." Make one mistake and you know the feedback is coming. Each person or manager has his or her own style. This is no different at our company. Even though I am only a team leader, I believe in the idea that for every single negative feedback you have to give a team member, you say ten good things about the work they do and constantly let them know they are appreciated. This way, when a mistake is made the team member accepts the coaching. It isn't seen as a way to beat someone down but as a way to help him or her improve. This method requires more work and time invested initially. What you will find is a lot of satisfied employees happy to come to work every day and eager to accept a new challenge.

Something else that I have learned along the way is that a truly great leader doesn't have to be an expert on any subject. Great leaders need to know about each subject, but first and foremost have to surround themselves with experts and then motivate those experts to believe in their abilities and have confidence in themselves.

When I first began this book, I started to ask some of my friends and people I respect to write. They all manage the way I would want a leader to lead. As I began to get these stories, I realized that I had asked the right people.

"One of the big lessons I learned is that it is OK to make mistakes, unlike many employers. The essential part of making mistakes is to learn from them and make improvements."

James Basham: Assistant Manager

Life with Toyota

May 8th, 1989, was a great day for my family and me. This is the day I started working for Toyota Motor Manufacturing. At twenty-five years old, a former farmer and construction worker, I didn't know what to expect. There was no way I could know all the opportunities and skills that I would learn over the coming years with the company.

I started as a team member, where I learned many of the core values of the company. One of the big lessons I learned is that it is OK to make mistakes, unlike many employers. The essential part of making mistakes is to learn from them and make improvements. Many times we try to make the big improvements, but in life we must try to make the small incremental improvements. These small improvements over time will lead up to great things.

One of our first big cost challenges that I can remember was a vehicle cost reduction activity. Each team member was given the opportunity to offer ideas to the design of the car to reduce the vehicle cost. Our goal was to reduce the vehicle cost to the customer by $100.00 for the next new model change. I suggested an improvement to change the engine decals to a molded part. With my idea and many others, we not only met our target but exceeded it.

This kaizen mind that I have developed has stayed with me, not only at work but also in my home life. I have learned problem-solving skills and interpersonal skills which have allowed me to venture out of my comfort zone. I now work with the kids in the youth program at my church and at church camp helping them learn and grow. This has helped me to achieve my goals at Toyota, working my way from a team member level to assistant manager.

Being a team member for this company is a great privilege and a true blessing.

> James's current position has him being the assistant manager in charge of the new model Camry launch. One of the first times I met James, I was an acting group leader for a group in chassis section. We were having a parts fit issue that was causing added work to everyone. He was out on the floor supporting us and helping us run the line to ensure that quality was 100% for our customers. I have been a fan ever since. —Tim

"When you truly understand another person, then you can work with them at a higher level. This is the level we seek to obtain at TMMK."

Mickey Payne: Group Leader

I have been employed with TMMK for nearly 15 years. One of the pillars of the Toyota Way is "respect for people." This is what our company is founded on. At TMMK we operate under this philosophy, and it is one of the reasons we have been so successful. I am a member of the Assembly Diversity Steering Committee, and one of my goals is to understand the differences in our team members and how we can use their strengths to make our company more successful.

One day I had most of my group in working on a Saturday, getting ready for a major change on our line. I had been their group leader for several months and thought that I knew my group well. I decided to order pizza for the group for lunch to reward them for all of their hard work that day. I had many team members in that day, so I decided to order several different types of pizzas. After the pizzas arrived, I went around the group making sure everyone had gotten enough to eat. At this time, I noticed a team member that was not eating anything. I approached her and asked her why she didn't get anything to eat, and she explained to me that she could not eat anything with pork due to her religion. You can imagine how bad I felt for not having a better understanding of her religious beliefs. I offered to get her something else to eat, but she had already brought her own lunch. She came prepared because she was used to things like this happening. This experience made me want to be a better group leader. I began learning as much about my team members as they were willing to share with me. I soon took the opportunity to join the Assembly Diversity Committee and have valued my time with this group.

I had always been a believer in the teamwork concept prior to working at TMMK. I grew up playing all sorts of sports and learned at an early age that working together, as a team, will help you reach your goals. I thought I understood that concept completely prior to working for TMMK, but I was wrong. My time here with our company has taught me that to be a great team, you must understand each team member. When you truly understand another person, then you can work with them at a higher level. This is the level we seek to obtain at TMMK. As a member of management, I know that it is my responsibility to practice this philosophy every day and to teach it to all that I can. I believe in the company that I work for, and I will continue to help lay the foundation for the next generation of TMMK employees so that our company can continue to be successful.

> Mickey is one of the most respected group leaders I know. He is a true leader. I appreciate his story about diversity. It was a great addition to our book. —Tim

"It does not matter what your title is; your opinion is expected and respected."

Fenella Smith: Specialist, Talent Management

When I started working at TMMK in October 2004, it was obvious my first week that I was working for a company like no other. My first assignment was as human resources teams representative supporting the paint shop. Team members in paint taught me what it meant to be a good TMMK citizen and why it mattered through modeling the way and demonstration. Furthermore, team members in HR taught me how to connect with team members in meaningful ways. My manager, Myra Ridley, and assistant manager, Denise Hord, taught me the importance of building sustainable relationships and partner-ships based on common goals and objectives. Relationships are HUGE at TMMK. When people trust and respect you, they will go out of their way to help you succeed. This was a priceless lesson for me to learn during my early years. In fact, the les-sons inspired me to start TMMK's second business-partnering group, Toyota African Americans for Success (TAAS), which was created to bring together like-minded team members who are Thinking Beyond Your Current Circumstances. This group would have never happened without the tenacity and dedication of Paul Newby, specialist assembly; Shawnita Walker, specialist paint; Shavonda Jackson, specialist plastics; and Keith Jones, manager of diversity. Again, each individual taught me the importance of being a better student and steward.

Back in 2006, when I started working in organizational development, I quickly discovered that working with TMMK top leadership was like being in a classroom. While my student status started back in 2004, the learning was delightfully different. During this time Pete Gritton, vice president of HR at TMMK, just started splitting his time between TMMK and TEMA. Pete is the ultimate team member advocate and is dedicated to making Toyota a great place to work. Hiro Yoshiki, vice president of general administration and senior advisor at TMMK, was still working in the plant and had tremendous positive influence on how we did things in organizational development (OD). Hiro is an excellent teacher. His approach was always methodical and easy to understand. During this time, Craig Grucza was the assistant general manager of human resources, April Mason was the manager for organizational development in addition to other areas, and Mary Robey-Singer and Mary Diggins were both my assistant managers during my time in the group. Lee Thomas, specialist, was a peer, and one of the kindest people in the plant to work with. Lee taught me how to be a better team member by demonstrating his dedication to all team members. I am taking the time to name names because I do not believe people fully understand to what degree their actions can have a positive impact on people who are in their environment.

While it is appealing to believe we do things on our own, the truth is we never really do. I am a student at Toyota and all team members take time to teach me, in small meaningful ways, how to become a better student, a better problem solver, a better leader, and a better TMMK citizen. Simply listening to the way ideas were exchanged, what things needed to be considered, what type and kind of nemiwashi needed to occur, how data needed to be presented, what kind of kaizen was expected, how to effectively use teamwork, and how respect is always expected was enriching. Never in my career had I worked in an environment where every opinion mattered and you are expected to engage in the process. It does not matter what your title is; your opinion is expected and respected. During my one-year dispatch assignment (March 2008–April 2009) at TEMA in talent management again, I had the privilege to work with Pete Gritton, who again gained more respect and admiration for the way he is leading our company during these tough economic times. I also had an opportunity to work with Mike Price, general manager of HR TEMA; Tim Ward, manager of talent management; Matt Long, assistant manager of talent management; and a host of other wonderful people at TEMA.

Again, each person taught me the value of living the Toyota Way, how to engage in TPS, and the importance of becoming better at problem solving. The purpose of HR management in Toyota is to maximize the performance of the entire corporate body by constantly maximizing employees' abilities to think, develop, and act, aiming to achieve the company's perpetual prosperity. After reading this essay, you might be saying, "Why did she name so many people? I thought this was supposed to be about her story." The point is, it is not my story; it is our story. My growth and development have been enhanced because team members at Toyota continue to allow me into their classroom.

> I can honestly say that getting to meet Fenella has been a terrific surprise from this book. I am happy to call her my friend. —Tim

"I remember one of the most important lessons I was taught by one of the Japanese coordinators was to go around and speak to every one every day."

Gary Cummings: Assistant Manager, Plastics Department

I started out my career on the Chassis 1 line, a week after my hire date February 20, 1989. That first week was many classes on safety, TPS, and just everyday items we needed to know about this place. My first job was installing fuel tanks; during my first few months on the job I did the same job every day for approximately three months, until I started learning other jobs in the team of four, in which everyone in my team got to do every job. It wasn't long that I knew many jobs on the chassis line. This was pretty cool, I thought, coming from a home builder background—and the Army before that. It was cool because I thought I would not like it; I had in fact become very curious about how many jobs I could learn, and is everyone's boss in this company as nice as my boss? I have since worked for many supervisors, and they all give 100% and respect the team members to the highest level.

I took team leader classes, to prepare myself for the next level, and quality circle classes, to become a circle leader, which made it easier teaching and coaching my team on problem-solving issues in our area. I was soon promoted to team leader in the same group. I found that it was a very hard position, training team members and doing other projects for the group leader. But it wasn't long before I found this position a great place to be.

After that I took group leader classes, hoping to someday become a group leader. That day came and I really got to move around. I was a group leader on two chassis lines, three trim lines; it seemed that when I would go into a new group, there were issues and a lot of kaizen that had to happen. Then, when we were running well, I would get moved again. What I didn't realize was that the assistant managers and managers were grooming me for the next position: safety group leader; not just to cover Assembly 1 but both Assembly shops 1 and 2. This seemed to be a lot of people to cover, but with the help of a couple of co-op temporaries and the management team we seemed to be able to cover it all.

Then the day came that I got a position as assistant manager in the Assembly Plant 1, and I welcomed the change. As assistant manager, I was lucky enough to cover all three areas at one time or another: trim, chassis, and final. Now this was a really good experience and it was great to have such great managers to have the patience to guide me through all of this development I was getting. I made several mistakes but not many times did I make the same mistake twice.

I love the things that the Japanese have taught me, and I hope I pass them on for the rest of my life. I remember one of the most important lessons I was taught by one of the Japanese coordinators was to go around and speak to everyone every day. At that time I thought he was crazy; that was over 90 people, how could I possibly talk to everyone every day? So one day he just grabbed me by the arm and we together walked through and talked to every team member in my section; it took about 45 minutes. I thought, now he will understand that can't be done every day. The next day he came by and grabbed me by the arm, and here we go again; this time it only took 30 minutes, and I learned about some issues that were going on that I had not known about. The next day he was there again and this time we got through in about 20 minutes and again I learned about more issues. Before he quit coming around, we got it about 15 minutes and I didn't want to quit; I was learning a lot from the team members that I would have never known if he had not made me walk through. I do admit I don't walk through and talk with every team member every day, but I do get around a couple of times a week.

Other things that we all learn are as follows:

- Always kaizen make it better: Kaizen—continue improving.
- When problem solving, always ask why: Don't stop asking why until you can't answer any more, then you will solve the real problem—root cause or the start of the issue.
- If the student hasn't learned, the teacher hasn't taught.
- Join a quality circle: A quality circle is when a group of team members gets together to work on a problem. See how the diversity of many team members can solve a problem better than one person.
- It is OK to make mistakes, as long as you learn from them and make sure you don't have any repeats of the same—improve yourself with each mistake.
- Save a penny 100 times, and you save a dollar, so when you are building 900 vehicles a day and you save 1 cent per vehicle you have actually saved $9.00 a day, which is approximately $2,160.00 a year. So any savings you can do is worth doing.
- Standardized work: Everything needs to have a standard, and standards need to be followed if we are going to stay competitive. When every one does the same job the same way, it is easier to see out-of-standard issues, or if a suggestion is truly better.
- Just in time: Only make what you need, when you need it, and the chance of it getting damaged will be less likely; also if you have a defect, it will be a lot easier to control.

Now I am in the plastics department, and have learned a lot more about why this company is so great. I was able to bring a fresh set of eyes into another shop, new ideas that I had practiced in assembly, and get a chance to see other ideas that could have made things better in the assembly shop, even though there are so many things that the two shops don't have in common. One thing for sure is TPS does work everywhere.

I feel very lucky to work for a company that believes in the development of its team members, also committed to ensuring each team member has the opportunity to work in an environment where everyone treats the other with dignity and respect. A company that will not give up on anyone that is willing to work here. My family and I have felt very good about this company since I was hired on in 1989, and now my son is employed here; even though he is serving in the U.S. Army and doing a tour in Iraq, we all know he has a position waiting for him when he gets back.

I could never say enough good things about such a great company as Toyota. But I do want to say, "Thanks, Toyota."

> I was a new team member when Gary would get moved around from group to group. If a group was struggling, he would go in and fix it. I appreciate his willingness to write for the book and share some of his insight about Toyota. —Tim

"I started out working on the assembly line, and had the awesome opportunity to travel to Japan and learn 'hands on' from the best."

Mike Goetz: Assembly 2 Assistant Manager

I was going to college and received the phone call I was hoping for from Toyota. I didn't know much about the company but did know the Japanese business model and philosophies were considered one of the best in the business world, so it was a "no brainer" for me to move out of the dorm and begin my Toyota career.

I started out working on the assembly line, and had the awesome opportunity to travel to Japan and learn "hands on" from the best. I was fortunate to have repeat visits to continue my education of the Japanese systems. I'm scared to fly now, so I'm glad that's over.

Once I became a teacher in the Toyota system or my official role as a team leader, I was responsible for teaching five team members how to perform their jobs. I was very proud to have moved up and was awarded much more knowledge by the company. My biggest memory and skills with this role was the way in which we teach people and how simple it was. I couldn't imagine doing it effectively in an assembly-type environment any other way. We called this JIT or job instruction training. I carried this outside of work when working with kids in sports and other activities. The other important item I learned in the challenging environment of building a car in fifty-five seconds was the way in which thousands of people all working in different areas, doing different jobs could allow the continuous movement of this vehicle down the line without tons of things going wrong. One of the items I grasped from this was the "best people" were working in the "best system." That is corny in my last sentence, but team members (the nickname for our employees) are given the responsibility and expected to control what happens to them during that fifty-five seconds of building that vehicle. The team member has the power to decide for the company if something is wrong and call for help with the andon (rope that alerts team leader to come help). I have been in other manufacturers' plants, and that just does not happen. I can't imagine thousands of people working independently, unsupervised, and holding the reputation of the company in their decision making. That is what happens every day at Toyota plants. It is a culture that exists within our four walls. The culture that exists starts from day one and the continuous education that occurs through the years to force us to use our brains to observe, evaluate, and solve why something isn't correct.

The last little thing that I like is how many of you have....

Driven home from work and walked up to the wrong house? Before you switch lanes on the highway, you look and use your turn signal; when you approach an intersection, a yellow light means caution and red light means stop. Airline pilots check certain items before takeoff and landing. These types of behaviors allow us to repeatedly do things without much effort and minimal failure. If we don't do these items I listed above, the outcome could result in injury or be potentially fatal. Toyota has mastered the method of establishing standards and identifying very quickly when standards aren't followed, and forcing us to react. Toyota makes mistakes, but we are expected to understand "why" and take ownership to find out "why" and ensure we learn from them and prevent reoccurrence. We don't just say that; we do it.

The neat thing about Toyota's success is that it is "inside of us and what we know," not what you see when touring or allowing our competitors to come in and visit. What they see is our robots and equipment and other tangible items are not what separate us from them. It is the way in which Toyota team members are taught to think and the standards we follow.

Yosh!

"Toyota team members' trust for the company is that high; in return, Toyota gets the best that the team members can give them every day, day in and day out. We care for this company like it is our own, and we are treated and act as if it is ours."

Mark Crupper: Group Leader

My story at Toyota started on December 3, 1993. My first child had just been born a few months prior, and I was excited when I got the offer of employment from Toyota for a good job with benefits to help provide for my family.

I can remember people asking me before I had started, what type of work I would be doing, and I would say, "I'll push cars from one end of the plant to the other if that's what they need done."

Fast forward fifteen years and two promotions, I had developed a pride in working at Toyota and being a part of the "family." I say "family" because you spend more time with your work family than you do with your real family, and also because working here feels like a family. For the most part, you and 7,000 other members are working toward goals to be successful on an individual and group level and the interdependence of each other.

Have you ever seen a trust exercise where a person will stand facing away from another person or group of people and that person has to fall backward and trust the other person or people behind them that they will catch them? Toyota team members' trust for the company is that high; in return, Toyota gets the best that the team members can give them every day, day in and day out. We care for this company like it is our own, and we are treated and act as if it is ours.

Hard work but doing it safely, treating EVERYONE with respect, continued improvement, and building in quality are what I feel separate us from other companies. Toyota does not have just good people scattered throughout the organization that make it a good company. Toyota is a good company with good people and family-type camaraderie. That's why I'm proud to work here.

> Mark is an excellent group leader with a bright future at TMMK. I would be proud to work for him someday. —Tim

Chapter 11

Our Development: Special Projects Stories

At TMMK we all take great pride in ensuring that our customers get great-quality cars. We do this by first building in quality at our processes. We have key team members that help look for quality issues so we can place quick countermeasures to resolve any issues that occur. We also have team members who help with safety or kaizen activity. They are the team members or team leaders that were just building the cars. Who better to support fixing the issues that arise? We realize that providing a safe work environment allows our team members to focus on quality.

All these team members have a great desire to be a part of the solution. These jobs are considered to be special projects. This allows us to develop our workforce. The special projects team members sign a deal to work in that area for 2–6 years depending on the position. After their time is up, they go back to the line or in most cases they get promoted. This is a system used to make each of us stronger to the company.

The following are stories from those key people who help the team members on line be successful in their goal of placing our customers first.

"There is nothing between you except air and opportunity."

Dwayne E. Lumpkins: Group Leader

I was selling cars for a Ford dealership in Winchester, Kentucky, when I finally got the call to come to Toyota in December of 1996. I had been trying to get hired for about three or four years when they finally called me and made me an offer of employment. I felt like I hit the LOTTERY when they offered me a position at TMMK.

I started on December 2 and had to go through eight weeks of training before I could actually work on line by myself. I could not believe all the training and exercises that I had to go through to become a Toyota employee. What company invests their time and money to let you exercise and get in shape to work on the assembly line? I could not believe the detail and the commitment to safety and quality that they taught us. In my new hire class, we were asked what our five-year plan was, and most people said they wanted to go to first shift. I told them I wanted to take classes and become a team leader, and most of the class laughed and said good luck with that.

When I finally got to work on the assembly line, I decided to prove them wrong. I got involved in quality circles, safety task force, and delta "S" task force, and applied for team leader classes after my first year as a team member. Believe me: I was involved in everything I could that would help me gain knowledge. I spent two years on line and was promoted to team leader in January of 1999. Most people told me that it would take four or five years to get promoted, but I didn't listen to them.

I had never worked for a company that wanted their people to succeed as much as this one. I had been a team leader for about two years when I got the opportunity to go to special projects and work in different areas and lines in the plant. We received eighteen weeks' training for the project group I was in. We were called SMK, which means "smooth motion kaizen." I could not believe that a company would invest that amount of time in me and believe in me to help make the company a better place to work. What I didn't realize at the time was that this was the Toyota Way.

You have a vested interest in them, and they also have a vested interest in you. I had a manager in the company ask me one morning what my five-year plan was now that I was in a project group. I told him that I wanted to learn as much as I could and try to become a group leader. This time, I was told hang in there—you CAN do it. Believe me when I tell you SMK was one of the best things that ever happened to me. I got to see firsthand how things worked from the management side and from the team member side of things. This was a real eye opener for me because I now started to understand why certain things were the way they are. SMK helped me develop as an employee and as a person. I was taught how to do many things that I could use at work and also at home. I was very impressed with the people who I worked with, the level of skill they had, and the commitment to the company they had. I can honestly say that I learned more in SMK than I even thought I would.

I started group leader classes in late 2005 and was promoted to group leader in July of 2006. So what I am really trying to say here is this: "Never let anyone tell you that you can't do something." There is a saying that I tell my team members now: "There is nothing between you except air and opportunity."

Dwayne is a well-respected group leader. He works in conveyance in Assembly 2. I remember when it was announced that I was going to safety, he told me that I would learn more about the company than I thought possible. He said I would see the big picture. It didn't take me long to realize what he was talking about. —Tim

"I also remember saying, 'This is almost as good as winning the lottery.' He didn't quite think it was that good. To me, it was the opportunity I had been waiting to get for quite some time."

Donald W. Tyra: Group Leader

It was in the fall of 1993 that I received a call from Toyota that I had been selected for hire. I remember it very clearly as I was working two jobs at the time to try to make ends meet as well as working out of my garage at home. One of those jobs was working for my father as a roofing supervisor, and the other was working as an auto body repair technician (which was also my extra work at home out of the garage). Times were pretty rough as I had two children then and one on the way. I remember I was getting ready to start painting a yellow Chevrolet Nova in my garage and my wife (now of twenty-three years) called out to me to come get the phone right away. It was about 7:00 p.m. and I didn't want it to get dark before I was done, so I was a little frustrated. Frustrated, that is, until I answered the phone and the man said, "I would like to know if you are still interested in working at Toyota in Georgetown." I said, "You've got to be joking, of course I do ... I mean, are you serious?" He explained what I needed to do in the upcoming days. I also remember saying, "This is almost as good as winning the lottery." He didn't quite think it was that good. To me, it was the opportunity I had been waiting to get for quite some time.

I was actually hired in on November 22, 1993, the Monday before Thanksgiving. I worked three days and received the Thursday and Friday off with full pay! Wow! I also remember how overwhelming the plant was the first time I came in, and yet I was made to feel very welcome. Christmas came very quickly that year, and once more I was overwhelmed by not only the company's practice of paying me for shutdown but also the generosity of the people. I had been training on first shift and this was not even going to be the shift I was on, but Sammy (first shift Chassis 2 team leader, or TL) treated me as if I was one of his own when he handed me a gift bag. This seemed to be the way everybody at the plant worked. I later learned that this was just part of the greater Toyota philosophy of how to treat people with respect. The temperament of the team members as well as management was one of patience and diligence. One of our basic-training theories at Toyota is "If the student hasn't learned, the teacher hasn't taught"; this way, the student is never the failure. It is the responsibility of the trainers to make sure that every team member knows the different nuances of their job in order to build a quality vehicle.

I have been very fortunate to be able to work several different jobs at TMMK, from chassis section to CART (communication analysis repair team), to CART TL, to a trim section TL, to the pilot section, and even to the Indiana facility (SIA) as a pilot TL rep for the chassis section at start-up. I am now currently a group leader in the Assembly 1 section of chassis. The thing I have always been overwhelmed with here is how people respond to the Toyota Way principles. These principles are not just a way to run a company; they work in every aspect of life. Two of the key points of the Toyota Way are "respect" and "teamwork." Team members here react differently because they are valued and respected for their contributions.

I was told when I hired in that I wasn't hired to build cars; I was hired to solve problems. Problems that, when solved, would contribute to the overall success of the individual and the company. Interestingly, when you employ these ideas to your work ethic it has a lasting effect. The people you work with see things in a different light; as you begin to solve issues and help others, you begin to build relationships that go far beyond your work. Relationships that help people understand and trust one another on a level far greater than any other workplace I have ever been involved with. A personal high point in my career here has come each time I move from group to group. Team members genuinely express their appreciation for a job well done.

The greatest expression of how genuine a relationship at TMMK can be came to me at one of the lowest points in my life. My mother passed away in August 2006. I had recently changed groups when this occurred, and almost every single team member and team leader from a group that I was no longer associated with showed up, after work, on a Friday at the funeral home. Not only did they show up but also they handed me an envelope with cash they had taken up to help out as well. (My sincere overwhelming appreciation goes out to the MA110 Trim Group as well as the pilot group.) TMMK also sent flowers, food, and a company representative to the funeral home. At a time when I was down, my extended family at TMMK went above and beyond, as they always do.

The team members at TMMK are a testament to Toyota's commitment to them as people respected for the way we continue to daily work through problem after problem to make the best car the world has ever seen.

> Don is the group leader of Chassis 3 in Assembly 1. He is responsible for parts such as the bumper and also that the brakes get filled with fluid. This is not an easy line, but he does it well. He also helped set up SIA (another plant), which also builds the Toyota Camry. —Tim

"Every day, I see it in action, and it never ceases to amaze me at what a group of driven individuals can accomplish."

Kevin Dunn: Group Leader

I was born and raised in the mountains of Hazard, Kentucky, and unfortunately the options that one has for a steady job are limited. So, I moved away and attended college at the University of Kentucky, where I pursued a degree in biology in hopes that one day I would be able to work in the medical fields.

Oh, how quickly things change.... In 1996, during my senior year in college, I applied for a temporary work position at Toyota Motor Manufacturing in Georgetown. I figured that I could make enough money to pay the bills and to support myself as I was about to complete my degree. In June of 1996, I was hired into the temporary workforce at TMMK. After a very short while, I gained a new understanding of the concept of teamwork and what it meant to be a functioning part of a group. For six months I worked on the trim assembly line; the longer I was there, the more I developed a desire to be a permanent fixture inside the plant. So, knowing that there was a high demand to be employed by Toyota, I decided to go ahead and submit my application for full-time employment.

In the fall of 1997, I received a much-anticipated phone call that I had been hired, full-time. My first day was January 19, 1998; this is the day that my life would forever change. Time seemed to go by in a blur, and I was fortunate enough to be able to work in many different groups along the way. Because of the many diverse people that I was working with and because I was exposed to so many new ways of thinking, I was starting to understand the concept of the "Toyota family" and how that notion was one of the strengthening fibers that promotes teamwork. It didn't take long as a team member to realize that I had found my niche. But, I wanted more.... I wanted to be in a position where I could help other people in the way that so many others had helped me.

From that moment, I pursued a promotion; I wanted to become a team leader. I wanted to be the one that the team members came to when they had a process-related problem. In 2001, my desire to team lead became a reality. I worked vigorously to soak up as much knowledge about Toyota and its values as I possibly could. For two years, I worked as a production team leader in the chassis section. I was active in quality circles and spent most of my time trying to engage team members in activities that would promote team building and group cohesion.

In 2004, I had heard that there were openings in a special project group called "pilot" that was responsible for coordinating and launching our model change activities. I submitted my application and was accepted into pilot. I didn't know what to expect when I first reported to my new group. The room was filled with computers and drawings, and I wondered what I had gotten myself into. Fortunately, I was again welcomed into the group with open arms. They worked to help me understand what was expected of me on a daily basis. After a few months, I became proficient in using the computer programs required to generate data required for a major model change.

When I finally thought that I had found my comfort zone, my group leader called a meeting with all of us involved with the Camry major model change. What I was about to hear would change my understanding of "one Toyota" from that point on.

My group leader informed me that I would be sent to Japan for three and a half weeks to help with design issues for the 2007 model. For a moment, I think I may have passed out. Remember, I was from a very small town and hadn't been on a plane for more than an hour or two at the most. In April I boarded a jet and set off on a thirteen-hour flight to Toyota City, in the Aichi Prefecture of Japan. I had no idea about what to expect when I arrived.

Much like most everyone I had worked with at the plant in Kentucky, the Japanese were quick to welcome me. I was paired up with representatives from Toyota plants in Japan, Thailand, Taiwan, Australia, and China. You can imagine the difficulties I expected when I realized that I had to discuss design- and process-related issues with people who spoke minimal (if any) English. Once again, teamwork prevailed, and we were able to successfully move past that stage of the model change. It was evident to me that teamwork not only existed at TMMK in Kentucky, but also is a value that is taught globally.

The people that I met from all the Toyota plants around the world each taught me something new that helped mold me into the person that I am now. Now almost six years later, I still keep in contact with a couple of these team members. I have a coworker, and friend, that I met in Japan who was from the Australia plant that I still speak with on a regular basis.

Two and a half years ago in 2006, I was offered a promotion to group leader. Without hesitation, I accepted. Now, I supervise thirty-seven team members and am fortunate to still be in the position where teamwork prevails. Every day, I see it in action, and it never ceases to amaze me at what a group of driven individuals can accomplish. I am thankful to have had this opportunity and look forward to having many more years of working with my team ... "Team Toyota."

I will leave you with one thought. In speaking with my team members one day, I told them this: "Don't ever think that we are here just to build cars ... instead, look at it like this ... we make cars ... but we build people." That's what makes teamwork a true asset to Toyota.

> Kevin has been given many challenges in his group leader role and his role in the pilot section. He always reaches his goals and, most importantly, helps his team members to reach their goals. —Tim

"Needless to say, by the end of the day I had learned
that there are a lot of people out there trying to do
as Toyota is doing."

Anthony Morgan: Group Leader

I came from a small community in eastern Kentucky where
family values and strong religious beliefs were taught early.
Where small farms were, for most, their only source of income
and starting a day's work was getting up at daylight and working until the light was gone. Having six brothers and three
sisters, I learned early that if I wanted anything extra, I had to
work and earn my own money for it, so, at the ripe old age of
eleven, I started my first job as a janitor's helper at my elemen-
tary school.

I continued to work all through school, and during most
summers I held two jobs. Like most teenagers around that
area, I couldn't wait to graduate and get out, and that's what
I did. I joined the Marine Corps. I spent eight years there,
through Desert Storm, with a wife and two small children; and
looking at another deployment, I decided it was time to look
at other ways to support them. Being in the service instilled
a deep sense of pride within me. Pride for not only my country but also pride in myself as well. After I left the service, I
started work for the federal prison system as a correctional officer, and it was while I was working there that Toyota called.

The money they offered was better, and I knew I wouldn't
be locked up with 120 inmates every night, so I was very
happy to take their offer. I guess the best way to explain it
is that I had been looking for a place to call home for the
past fifteen years, and it didn't take me long after starting at
Toyota to know that that was just what I had found: a home,
a place I could work and be proud of, that would support
my family and give us a good life, a place that took care of
their employees. A place that not only preached the same

values and principles I was raised on, but also taught them and lived them.

I started work at Toyota in the winter of 1994, starting up production on second shift in Plant 2. I moved to day shift in two years, and then tried conveyance for a year before moving back to chassis.

I took team leader and went back to second shift in chassis. I then took a position in the "tool crib" as an air tool repair specialist, and it was there that I got to see a lot of the technical side of producing vehicles. While in the tool crib, I received a promotion to group leader and am currently serving as group leader in Plant 1 Chassis 1.

I have held positions from instructing marksmanship and making explosives in the Marine Corps, from guarding prisoners to being a janitor; I have worked for fourteen different employers, and I can honestly say this is the best company I have ever worked for. This shows in our team members, team leaders, and management.

Toyota is not just another job for most of us; it is not just a means to reach something better. It is the "something better." Toyota continues to prove this during these hard times that are before our nation and in our industry, and I know that Toyota will hold fast to their beliefs and values during these times.

The Toyota Way is sought after and taught at other companies around the globe. I had only been working for Toyota about six years when I got to hear about some other companies for myself by the employers of those companies. It was on a fishing trip with my brother-in-law, who had visitors from New York and Pennsylvania down to Kentucky.

Each member was a high-ranking official or owner of companies from those two areas. Needless to say, I was left out of most of their conversations throughout the day and felt as though I did not belong in the same boat with these guys, until one finally turned to me and asked what I did. When I replied that I worked for Toyota of Georgetown, their attention turned to me. The way they asked about Toyota, their eagerness to learn, and the remarks they made—suddenly I felt like I was head of the class.

They could not understand kaizen, continuous improvement by our team members, or how we could allow a team member to make changes to his process. Needles to say, by the end of the day I had learned that there are a lot of people out there trying to do as Toyota is doing. Incidentally, I caught more fish than all of them put together.

Hard work is a part of any factory, and Toyota is no different, but Toyota allows people to use their minds; not just allows, but encourages it, relies on it, and depends on it. We all work as a team, and this is what makes Toyota great and makes us all successful. It is on these grounds that Toyota stands, and it will be on these grounds that Toyota will still be standing when the smoke clears and the fires die.

> I have known Anthony for a long time. He is now the group leader on engine line in Assembly 1. He does an excellent job. —Tim

"I have always had success with this style and think that the time I spent in the safety department has development me into a stronger leader for this company."

Mike J. Perkins: Group Leader

It was the fall of 1996. I was nineteen years old, and my wife had just had our first child. Life hadn't turned out exactly how I had planned it, but now with a young daughter and wife that I had to support I needed to grow up fast. My father, who had worked for Toyota since 1989, told me that Toyota was going to start hiring team members from the temporaries that worked in the factory. I had been trying for over a year to get hired, and the waiting process at that time was 3–5 years. I thought this was my opportunity to get in the door and prove what I was capable of.

I worked as a temporary for one year and in January of 1998 was hired on as a full-time team member. Even as a temporary, I was treated extremely well. I never felt that I was any different than any of the other Toyota team members. They treated me as if I was an equal to them and never made me feel out of place. This was a surprise to me because at this time I knew very little of Toyota's values and that respect for people is one of their key pillars to success.

Once hired full time, I was asked what my ambitions were over the next few years. I told them I wanted to be a team leader within two years. I got the "young kid ambitious speech" that you would hear from time to time, but got a lot of support from my group leader and team leaders. I was promoted to team leader in October of 1999, just under two years.

As a team leader, I learned a lot about the Toyota Way. How Toyota prides itself on respect for people and continuous improvement. I saw how servant leadership helped to develop the people and leadership of the company, making everyone that much more valuable to the company's success. This was a philosophy that I really believed in and wanted to practice on a larger scale. I applied for special projects, and in 2001 was selected to join the Safety Task Force. This was an ideal place for me as now I would be looking after the needs of several hundred people.

Being involved in a lot of different areas around assembly, I could see a lot of the problems all the team members were having and hadn't been able to get resolved. The problem at the time was they didn't know how to fix the issues they had. I proposed to my group leader that if I had a team of individuals that could work on these problems how much we could accomplish. He allowed me to use three team members to work on these problems, and thus we started the safety kaizen team (continuous improvement). This team was built with all team members. They would take problems other team members had identified and correct them. Toyota gave us the training we needed to be successful, whether that was welding, pneumatics, or just ergonomic training in general. With all these tools we were able to better serve our team members, and in doing so we cut our incident rate in half in just over a year. By eliminating the safety concerns of the team members, it has elevated their morale and has gotten them more involved in the improvements.

> Mike was the safety representative before I came to Assembly 1. He was a tough act to follow. His commitment to safety has not wavered after he became a group leader. He has not had a recordable injury in any of his groups. —Tim

"I will always have a strong sense of responsibility to Toyota because of this, and because of many things they do to ensure my safety and long-term employment."

Brian Eddy: Team Leader, Early Symptom Investigation Team

In November of 1994, my supervisor at Delta Airlines came up to me and told me I was one of the unfortunate 1,500 people to ever be laid off by the airline. Their losses at that time had been in the millions, and the cost-cutting measures they had taken were not enough to stop the bleeding of money. I had seen the writing on the wall three and a half years before and had been going through the hiring process here at Toyota.

The uncertainty of my employment situation was nerve-rattling at the least, and I had set up to work at the local grocery store to make ends meet till I was able to get another full-time job. I called Toyota as I had weekly for three and a half years, but this time the response was different: "Can you come in for a physical, Mr. Eddy? You have passed your day of work, and we are ready to provide you with a full-time position contingent on the results of your physical."

I passed the physical, and from the time I was laid off till the time I walked in for my first day of work at Toyota, I was unemployed a total of a week. I thanked the Good Lord for the job, and the new challenges that lay ahead.

I will always have a strong sense of responsibility to Toyota because of this, and because of many things they do to ensure my safety and long-term employment. I went to an ergonomics convention for Toyota years ago. I was amazed in talking to many other auto manufacturers, aircraft builders, and so on that none of them had anything like our Early Symptom Investigation Program here at Toyota.

When I would talk about the Toyota Ergonomic Burden Assessment we use to rate jobs in the plant, they were amazed Toyota put that much time and effort into trying to find burdens in the processes, and that they had a specific tool to numerically measure this. This trip was eye-opening for me; it gave me tangible proof that Toyota goes above and beyond to provide us with a safe and stable work environment. They certainly aren't perfect, but in my book they are head and shoulders above the rest. I am proud to be part of the Toyota team, and I am thankful for the opportunity Toyota provided me.

> Brian is the safety rep for plastics. I had the opportunity to attend a class with him and was very impressed with his knowledge and commitment to safety. —Tim

"Now I am directly involved with improving the quality of our products."

Dean Weddle: Team Member

For me, the hiring process covered a five-year span. My reason for seeking employment at TMMK was merely out of curiosity. I had only recently been promoted at my current employer and was very content in my new position. I had a great supervisor and enjoyed working with the other employees. So when TMMK would call me for the next interview or test, I really just went for fun to see what would happen. Now, I have been a team member since October 12, 1998. Getting hired turned out to be a timely blessing.

When TMMK called and offered me a position, it turned out the company I was working for had only a month earlier informed me they were restructuring and downsizing. In other words, my position was being cut. So needless to say, I jumped on their offer and accepted the position. I had heard good things about working for Toyota, but I had never performed factory work before and had no idea what to expect. Prior to starting my job, they routed me through an eight-week training program that helped prepare me both mentally and physically.

My first job was working in Assembly 1 on Trim 1. This is the first line in assembly to start putting parts on the car body. It was an eye-opening experience to see how a car is actually assembled. I had a great team leader that encouraged—no, I should say pushed—me to use both hands when performing the jobs. I came in dominantly right-handed. Today I can use a tool to shoot just about any part with either hand. Learning this skill has greatly decreased the ergonomic burden of repetitive work. This skill has also helped me with home repairs. I used to have to twist in order to use my right hand. Now I just use my left.

At most companies, in order to change the type of work you perform you would have to change companies. At Toyota, they provide a transfer system that is based on seniority. If you become bored with the type of work you are performing, you can watch the transfer system and move to another job or another department. TMMK provides work in several different departments: assembly, plastics, paint, body weld, stamping, and power train. It is even possible to transfer off the floor into an office environment without losing tenure. I recently transferred from Trim 1 to a quality gate team. Now I am directly involved with improving the quality of our products. In this position, I have worked on just about every line in assembly, and this has provided me a better overall view of how each team member is like a gear in a machine. Teamwork is a must in order to ensure safety for our team members and a quality car for our customers. Working in the quality gate team has been a personally satisfying position for me. It feels good to go home knowing you are in a position to help ensure the quality of the vehicles to our customers.

Yes, you could say that I am very similar to the Hair Club for Men commercial. I'm not just a team member; I'm a customer. My first Toyota experience was when I traded in my car for a 4×2 Toyota truck. This was a great little truck, but as we became more involved with landscaping we decided to get a Tacoma 4×4. We then bought a used 1995 Camry. I could not believe how well it drove and road. I eventually gave this car to my mother and purchased a 2004 Matrix. This provided us with more room for hobbies and great gas mileage. We later decided we no longer needed a truck, so we bought a 2006 Sienna built by our fellow team members at TMMI. This van rides like a luxury car. We love it. My niece needed to replace her car, so we gave her the Matrix and just purchased a 2009 Matrix. The team members at NUMMI (New United Motor Manufacturing, Inc.) do a great job building these cars. The 2009 Matrix has increased power without losing any gas mileage and an even better ride. We helped our grand-niece purchase her first car, a used 2000 Echo. As you can see, we are now Toyota customers because of the quality and confidence they have earned. I have yet to have to trade any of my Toyota vehicles in because of performance problems. My extended family and I have been very pleased with the Toyota experience.

Dean Weddle works in the quality group in Assembly 1. He does a great job at ensuring quality gets built into each car. —Tim

"I meet and talk with many of our team members, team leaders, group leaders, and assistant managers daily through this job."

Tony Smith: Team Leader, Tool Crib

I was hired at Toyota on April 13, 1998; prior to coming to this company, I had only had two other jobs, and never a job at such a huge company. I'm sure a lot of people don't believe in destiny, but as I write this, and as I reflect on my jobs and life, I do.

My childhood was as most: my brother and I were raised by a single mother, and my dad's job had taken him out of state after their divorce, so we only really got to see him during the summer. My mom provided everything she could for us, but of course raising two boys really tested us financially. I learned early in life if I wanted something, I would have to work as hard as I could for it, because my mom could only afford the necessities. This was difficult as a child to understand, but little did I know how this would help me later in life.

I got my first official job at sixteen at our local grocery store in town; the owner of this store was an older lady whose work ethic was off the charts. Her then deceased husband and her had opened this store forty years before I started working there with a few dollars and a dream. They took this store from two to forty people within the forty years, never expanding too quickly, never ordering more than they needed, and always saving for a rainy day. She was the toughest lady I had ever met; she had one expectation, and that was that the customer was always right. Your job every day was to say hello to every customer, ask can I help you, and always tell them thank you. We sold service and dependability; if we didn't have it, we could get it; no was not an option. To say I learned life lessons from her would be a huge understatement. I worked there six years, all the way through high school and college, learning from a lady who taught me from her life's experience; thanks, Mrs. T.

I got my second job just out of college at an environmental engineering firm, referred to this company by a friend of mine who I had grown up with. I had to interview with the owner and president of the company; to say that I was terrified would not even get close to explaining it. His office had one entire fifteen-foot wall full of diplomas and certifications; he was a graduate of Stanford, with a doctorate from the University of Kentucky. I again had no idea at the time how much this man would mean to me in my life; he rewarded hard work and excellence, and he was not shy about telling you. He had built his business on his word and his ability to deliver what he promised. He started this company with two people in 1978, and by 1998 had built it to 178. I reported directly to him almost my whole eight years at the company, and I grew to completely admire this man; I still today, when making decisions, sometimes think about how he might handle the situation. He mentored me in business and in the management of people, and to that I am grateful; thanks, Dr. F.

In March of 1998 I got a call from a man saying he represented Toyota Motor Manufacturing in Georgetown, and did I have a second to talk. I had applied at Toyota three years prior because of rumors that the owner of my company was considering retiring within the next few years. This man explained that they wanted to offer me a job, I couldn't believe it. I of course accepted, and started in April 1998 on Final 1 Plant 1. I had never worked in manufacturing; so to say it took some getting used to would not even be near it. I spent three years on line, my fourth year as an acting team leader, being promoted to team leader shortly after. I have been given so much opportunity while at Toyota, smooth motion blitzes, takt time changes, Simplification 1 and 2, and 770N launch and 044L launch teams. I am currently in tool crib covering both Plant 1 and Plant 2. I meet and talk with many of our team members, team leaders, group leaders, and assistant managers daily through this job. I believe one thing through my daily journey at Toyota, and that is that it is our true diversity and the Toyota Production System that is our strength. We truly have people from all over the world working together on a daily basis. We are successful because not just one part of the company learns this system: everyone learns it, everyone uses it, and everyone lives by it.

This job has truly been a blessing to me, my wife, and my two boys in so many ways: financially, its benefits, and its learning experiences. I will forever be grateful for that phone call in March of 1998.

> Tony has one of the busiest jobs at TMMK. He repairs and maintains the air tools we use. As you can imagine, keeping up with all the air tools in two plants that can produce 2,000 cars a day is a pretty demanding task. He takes care of us on second shift. He is the perfect person for the job. —Tim

Chapter 12

Toyota Way Principles

There have been many books written about the Toyota Way principles. Most of these books have been written by professors and consultants and are very educational. To the best of my knowledge, the principles have never been explained from the perspective of the hourly worker or on a personal scale such as this from the people who apply the principles in their work life.

The principles are a set of guidelines that was created by management at Toyota in the 1930s–1960s. These principles are one of the tools used to create our work culture and promote a lean-manufacturing organization. If you want to learn more details, I suggest you read the other books written. Some of my friends and I are going to attempt this through our own personal experiences and thoughts.

Long-Term Philosophy

Principle 1: Base your management decisions on a long-term philosophy, even at the expense of short-term financial goals.

- ■ Have a philosophical sense of purpose that supersedes any short-term decision making. Work, grow, and align the whole organization toward a common purpose that is bigger than making money. Understand your place in the history of the company, and work to bring the company to the next level. Your philosophical mission is the foundation for all the other principles.
- ■ Generate value for the customer, society, and the economy—it is your starting point. Evaluate every function in the company in terms of its ability to achieve this.
- ■ Be responsible. Strive to decide your own fate. Act with self-reliance and trust in your own abilities. Accept responsibility for your conduct, and maintain and improve the skills that enable you to produce added value.

The idea of this principle is to ensure the long-term survival of the company. Below is a story written about a recent decision that our company had to make during this recession.

February 12, 2009

This would become one of the most historical days in Toyota history; it would also teach many team members firsthand how the company uses Principle 1 in its daily business practices.

For nearly the last year, our economy had gotten gradually worse. Gas prices had soared to $4.00 a gallon, which devastated everyday Americans' budgets. We all had to cut back on everyday spending just to pay for gas to get us back and forth to work. The rising gas prices threw the whole economy into a tailspin. We were already in the midst of a recession, but the gas prices made it even worse. The lack of spending slowed down production in all areas of the country. Many Americans around the country were laid off from their jobs. The uncertainty surrounded everyone, including our Toyota plants around North America. Most recently, we had to shut down production at our Texas plant and one plant in Indiana. These plants built our full-size trucks. With gas prices around the $4.00 range, no one was buying trucks. Once the lay-offs began, no one was buying anything. It took a while for the economy to catch up with the mid-sized car segment. In November and December, it began. Just like everyone else, our car sales were slowing down.

All the team members knew that things were going to change. We had it extremely well for years. Our company would give away cars for perfect attendance. Our pay was equal to the pay of UAW workers. We had all the benefits that all auto plants had. In my fourteen years, we had never had a situation such as this. We were assured that steps were being placed to ensure that no full-time team members would be laid off. Unfortunately, this would not be the case for our temporary workers. It has been very sad to see these good, hardworking people leave our plant, but they will have the opportunity to come back. The economy will eventually improve, and steps have been put in place to offer these temporaries back their jobs once it does.

Toyota directly provides stable employment for nearly 36,000 Americans. The company has taken unprecedented steps to ensure the long-term employment of these team members. February 12 would be the day when all 36,000 team members would find out what steps the company was taking to ensure the long-term employment of everyone. Each general manager would share a presentation and all the information to their team members. These presentations would explain the economy and all the factors that played a role in our current situation within our industry.

No one knew how it would be perceived or what to even expect. We had no idea what we were about to be told. What we heard was a very good plan by very compassionate people. These general managers had earlier in the year accepted pay cuts and no bonus. They gave of themselves before they asked us to give up anything.

Every possible step had been taken to avoid the meeting we were about to have. As the economy became worse, this meeting and announcement could not be avoided. We would find that our company was going to keep every worker. They would be offering a voluntary exit program. This program is in no way a layoff. No one would be forced to leave Toyota. Many more team members took this option than expected. Our workforce had been going strong for twenty years. Many were at the age to retire and saw this as an opportunity to do so. Each of those team members is missed.

We would also find out that many programs had been temporarily put on hold. Some will return. Others may never come back. Our bonus was temporarily put on hold. I hope that comes back. It will be a sign to each team member that things are improving. Of course, I would like to have my bonus, but I think of it like this: if I had to pick keeping my pay or choosing a fellow team member to lose his job so I could keep a few thousand dollars a year, I choose to take the pay cut for a couple years. We will come back even stronger than before, and when we do the company will reward us for it.

You may wonder why I share this story when discussing Principle 1. I do so because our company is losing money every day to keep us all working. Our leadership in Japan realizes that keeping us working will provide us with good, stable, knowledgeable team members when the economy picks back up. Our quality would sacrifice if we lay off workers. Our principles and values would be sacrificed if we lay off workers. Right now our company at the very top is making the decision to stand by their principles, and for that I feel very thankful.

"When the going gets tough ..."

David Cox: General Manager, Power Train

Like everyone who's been at TMMK for any length of time, I've had numerous occasions to witness the quality of our workforce and the "whatever it takes" attitude when we have problems. From pitching in to help a struggling team member, to the warm welcome shown our TMMI team members assigned here during their downturn.

However, never have I been moved—nearly to tears—like I was during the announcements of bonus and pay cuts.

I found myself responsible to stand up in front of over 1,000 power train team members, and tell them in addition to the overtime they had lost, they would also be losing their bonus and TIE payments for the foreseeable future. Needless to say, I was nervous. This was news that would dramatically impact almost everyone's income and way of life.

The response was nothing short of amazing. Not one negative comment. Not one of the 1,000 people acted negatively. In fact, after the meetings and over the next few days, I had many people stop me to say, "I don't like getting my pay cut—but it's the right thing to do to keep everyone employed, and thanks for taking your cut, too."

Anyway, when the going gets tough, the teamwork gets that much stronger. I couldn't be prouder of where I work and the people I work with.

> David Cox is the father of eight kids and the general manager of one of Toyota's largest powertrain plants. He is a very dedicated man with a lot of responsibility. I have great respect for him. —Tim

The Right Process Will Produce the Right Results

Principle 2: Create a continuous process flow to bring problems to the surface.

- Redesign work processes to achieve high-value-added,
  continuous flow. Strive to cut back to zero the amount
  of time that any work project is sitting idle or waiting
  for someone to work on it.
- Create flow to move material and information fast as
  well as to link processes and people together so that
  problems surface right away.
- Make flow evident throughout your organizational
  culture. It is the key to a true continuous improvement
  process and to developing people.

When we design a process at any Toyota plant, we try to
make sure all the work possible is value added. What I mean
by that statement is this: When our pilot section designs the
processes and fills out our standardized worksheets, the goal
is to have each step add value to the car. If we shoot or install
parts on the car, that time is considered value-added work. If
we are walking or picking up parts, it is considered non-value-
added work.

Any time that isn't adding value to the car is in need of kaizen. One tool we use to do this is with our process diagnostics
sheet. This sheet breaks down our processes to show value-
added work and wasted time. It shows all information from
the light reading to the walk time on each process. This tool is
invaluable to our success. A group called SMK (smooth motion
kaizen) is there to support us in this ever-improving endeavor.

One example of how this tool is used to improve Principle 2
is the idea that if you start a part, you finish it. When our plant
first began, we would sometimes start a part, but the next process down the line may actually shoot the part to the car. One

way we improved our quality is through finishing what we start. Another example of this tool is to help with safety. We note any noise and light levels to each process.

A few years ago, we began a line side simplification project. During this project, we began using kit boxes to deliver parts to the line. These boxes are placed in the car in most cases at the beginning of the line. These boxes have all the parts needed to build that car. Of course the team members confirm they are installing the right parts, but we found this method to be invaluable in building in quality at the process. This method also helped us keep our plant even more organized and clean.

The parts are sequenced in a kitting area. The team members have printed sheets that tell them which parts goes in each box. Using this method is just another example of our thinking minds in action to improve upon and further our knowledge on Principle 2.

Principle 3: Use "pull" systems to avoid overproduction.
- Provide your down line customers in the production process with what they want, when they want it, and in the amount they want. Material replenishment initiated by consumption is the basic principle of just in time.
- Minimize your work in process and warehousing of inventory by stocking small amounts of each product and frequently restocking based on what the customer actually takes away.
- Be responsive to the day-by-day shifts in customer demand rather than relying on computer schedules and systems to track wasteful inventory.

Have you ever noticed at a grocery store that the milk is on a small ramp and is loaded from the back? Of course, this is done to make sure the milk gets sold and used before it becomes outdated and spoils. We see our products in the same way. Of course, they won't spoil like milk, but we don't want a car setting in a lot so long that it becomes a part of the landscape either. Cars are made to be driven, not parked.

Toyota was challenged with this principle this year. We try to keep around a forty-day supply of cars in the field. What I mean by this is we try to stay forty days ahead of what our customers buy. We do not like having cars setting at the lots for a long period of time. During the economic downturn, our cars in the field grew to nearly ninety days; once this occurred, we reduced our takt time to slow down our line to make up the difference.

We also use a kanban system to help us track our parts. The last thing you will see in a Toyota plant is a lot of parts setting around or a warehouse holding parts until we are ready to use them.

One story states that during a visit to the United States in the 1950s, Mr. Toyoda was impressed with the way our supermarkets would stock food. It was treated as a "first in, first out" method. Japan at the time did not have this sort of system at their grocery stores. Mr. Toyoda would take this lesson and apply it to all the Toyota plants. Our parts would be delivered, as we needed them. This helped us avoid needing wasted space to hold parts until we were ready for them. Trucks are constantly moving in and out of our plant.

Following this method also helps improve our kaizen time. When you work on a pull system, there is no room for errors. As problems arise, they are noticed immediately. This allows for quick countermeasures to be put in place to avoid that issue from happening again. This allows us to be more efficient in our daily operations.

Principle 4: Level out the workload (heijunka). (Work like the tortoise, not the hare.)

- Eliminating waste is just one-third of the equation for making lean successful. Eliminating overburden to people and equipment and eliminating unevenness in the production schedule are just as important—yet

generally not understood at companies attempting to implement lean principles.
▪ Work to level out the workload of all manufacturing and service processes as an alternative to the stop-start approach of working on projects in batches that is typical at most companies.

Leveling out the workload is done in many ways. One way to do this applies to the cars coming down the line. For example, in Assembly 2 we have four different models that are built on the same line: the regular Camry, the Camry Hybrid, the Solara Convertible, and the Venza. Some models require more seconds to complete the process than others. If we built two Venzas back to back, the team member would be behind in the process because a Venza is harder to build. The parts are bigger, and in most cases some processes require the team member to climb inside the car rather than just sit in the rear door.

When those cars are separated going down the line, it allows us to better plan our processes without overworking the team members. For example, if we have a process that takes longer on a Venza, we can make sure the work is lighter on the Camry. When the cars are sent down the production line in a certain order, it allows for a smoother process flow for the team members. The company tries to follow this principle whenever possible. At times the Camry may be selling extremely well, so more Camrys may go down our line. When this occurs, then standardized work can be changed to accommodate the new heijunka.

Another tool we use to level out the workload is a tool called TEBA. We can break down all the major steps of a job and check all details including the push force of connections that need to be made and the number of shots required to shoot all the bolts on a particular process. This tool is used to give each job a rating. If a process is labeled a red process, then we know it needs immediate countermeasure activity to get it within our standards.

A final example of heijunka is our inventory. When other factories shut down for months at a time to reduce their inventory, we just slow down the line and reduce our processes. This ensures our job security and allows us to manage our inventory rather than let our inventory manage us. In these cases, the extra team members will work in special projects groups to look at ways to reduce our total cost. During the recession, some team members with fabrication skills were taken off line and worked on projects that we would have normally hired a contractor to do. The cost savings from activities like this allows us to keep our workers working and allows each team member to feel as if they are helping everyone keep a job. This is an important aspect to Toyota in creating mutual trust and servant leadership qualities to its team members.

Principle 5: Build a culture of stopping to fix problems, to get quality right the first time.
- Quality for the customer drives your value proposition.
- Use all the modern quality assurance methods available.
- Build into your equipment the capability of detecting problems and stopping itself. Develop a visual system to alert team or project leaders that a machine or process needs assistance. Jidoka (machines with human intelligence) is the foundation for "building in" quality.
- Build into your organization support systems to quickly solve problems and put in place countermeasures.
- Build into your culture the philosophy of stopping or slowing down to get quality right the first time to enhance productivity in the long run.

John Pfost: Team Leader

The Andon

Proud, nervous, scared, uncertain, confused, empowered, in control....

What do these mean? These are all some of the ways that pulling the andon can make a team member feel. In my earliest memories as a team member, I can remember instances where I felt all of these. The most prominent of them, however, was the sense of pride.

Pride, you may ask, why would you feel pride? Doesn't pulling the andon mean there is a problem? Doesn't it mean you did something wrong? While this may be true, those aren't the andon pulls I remember. I remember the ones where I noticed a defect. A wrong part installed, a part missing from the car. All those times I was there and helped control the quality of that vehicle. Someday soon, someone would spend their own, hard-earned money to own it. A car that would deliver families to their destinations and get them there safely.

That is where I find the pride in pulling the andon. Those are the memories I will always have with me.

Many years later I have a slightly different outlook on the andon. It is in no way a negative outlook but instead a positive one. It's been nearly ten years since I began my adventure here. I've spent the years learning and growing to have a better understanding of the Toyota Way. I have worked my way up to the position of team leader. I accepted this promotion with the highest level of honor and respect for the position that I am in. Now, all these years later, I'm no longer on the outside looking in. Now I'm on the other side of the andon.

As a team leader, I see the different uses of the andon. Sure I still feel a sense of pride, but now it's accompanied by a sense of honor. Now it's my duty to handle whatever the situation might be when I answer an andon pull. I can see how my past team leaders dealt with all the situations. They took ownership throughout my years here, every time they answered the andons.

One of the most important ways the andon comes into play is as a tool. Through the use of tracking andon pulls, we can see a lot about our line. We can determine which processes have the highest number of pulls. We can even narrow it down to specific team members to see if someone needs more training or if a particular process is having issues. It is one of the best tools we have when problem solving and implementing kaizen.

As I stand here today and put my thoughts into words, it becomes more obvious to me why this company is so successful. They provide us with the tools we need to improve the company from within. Through the use of a tool called the andon that is nothing more than a piece of rope, we are able to grow so much as individuals, as part of a team, a team called Toyota.

John is another one of my former team members who was promoted. John was one of the few people that I requested a specific topic to write about. I think he explained the use of the andon and the feeling generated from it perfectly. Thank you, John, for accepting the challenge. —Tim

Principle 6: Standardized tasks and processes are the foundation for continuous improvement and employee empowerment.

■ Use stable, repeatable methods everywhere to maintain the predictability, regular timing, and regular output of your processes. It is the foundation for flow and pull.

■ Capture the accumulated learning about a process up to a point in time by standardizing today's best practices. Allow creative and individual expression to improve upon the standard; then incorporate it into the new standard so that when a person moves on, you can hand off the learning to the next person.

Most of us follow a standard whether we realize it or not. When we wake up in the morning, everyone generally follows the same path. We have a way of doing things. Maybe this is because we aren't fully woken up yet or maybe because we get set in our ways and aren't really fond of change. In most cases, it is because we maximize our time to get the most sleep possible.

Toyota follows this same philosophy with standardized work. We maximize our time to get the most work done in the allotted time available to complete our processes. Each process has a standardized worksheet. We also have a job breakdown sheet for each job for training purposes. This sheet shows how many seconds it should take to install our parts. We have key points and the reasons for those key points listed on these sheets to help the line workers to better understand why the work is being done that way.

Don't take me wrong; while these sheets are a standard that everyone must follow, they are also benchmarks for kaizen. If a team member sees a better way, we are encouraged to talk to our team leaders and group leaders about an idea and in most cases are allowed to run a trial to see if it is better.

Once it is proven, both shifts and all the team members communicate the new way and sign off on the new standardized work. Once that occurs, then we have improved our process for everyone.

Having standardized work is not just about doing the same thing all the time. It is about improving the same thing all the time while working together as a team.

Principle 7: Use visual control so no problems are hidden.

- ■ Use simple visual indicators to help people determine immediately whether they are in a standard condition or deviating from it.
- ■ Avoid using a computer screen when it moves the worker's focus away from the workplace.
- ■ Design simple visual systems at the place where the work is done, to support flow and pull.
- ■ Reduce your reports to one piece of paper whenever possible, even for your most important financial decisions.

There are visual indicators everywhere at any Toyota plant. I am going to discuss a few.

The Andon Board

Every line has andon boards hanging from the ceiling. John talked about the andon and its uses. When a team member pulls the andon, a light gets lit on a board for the team leaders to see. Each andon is numbered. The andon board lights up the appropriate number so the team leader knows where to go to. This is the most used visual control we have at any Toyota plant. It allows for a quick response to countermeasure and deal with problems as they arise.

The Andon Light

Also, each process has a light that the andon rope is tied to. When the team member pulls the andon, then the light comes on. This is also a visual indicator for the team leader.

Floor Management Development System (FMDS) Boards

FMDS boards have all the key performance indicators on them so all the team members know how their group is doing and what the targets are. This is important because we are all a part of the same team and need to be able to see all the information.

Man Figures on the Process Boards

Each process board has a man figure on it. This man figure shows all the proper PPE (personal protective equipment) needed to run that process. For example, if a process has a part with a sharp edge on it, then Kevlar gloves and sleeves are required.

Principle 8: Use only reliable, thoroughly tested technology that serves your people and processes.

- Use technology to support people, not to replace people. Often it is best to work out a process manually before adding technology to support the process.
- New technology is often unreliable and difficult to standardize, and therefore it endangers "flow." A proven process that works generally takes precedence over new and untested technology.
- Conduct actual tests before adopting new technology in business processes, manufacturing systems, or products.

- Reject or modify technologies that conflict with your culture or that might disrupt stability, reliability, and predictability.
- Nevertheless, encourage your people to consider new technologies when looking into new approaches to work. Quickly implement a thoroughly considered technology if it has been proven in trials, and it can improve flow in your processes.

One portion of technology that we use on a daily basis is what we call a "pokayoke." A pokayoke is a fail-safe device that helps prevent nonconformances from passing to the next process. This system alerts the team member when a problem has occurred at the process through lights, buzzer, or line stop. We typically use the pokayoke system on processes that have critical fasteners involved, that is, front and rear seat bolts, tire and wheel lug nuts, and seat belt nuts and bolts are just a few examples of how this technology is used. Most pokayokes are used in conjunction with tool control boxes, which have torque control standards set up within them; these boxes are connected by cable to the pokayoke box, and to the tool that is shooting the critical fasteners. These standards consist of a low torque standard, ideal or cut torque standard, high torque standard, and the amount of fasteners (nuts or bolts) required for each vehicle. If at any point within the team members' process, any of these standards are not met, the tool controller will send an out-of-standard alert to the pokayoke system.

These technologies are essential to our success; they are proven technology that we depend on daily. Our team members' customer-first philosophy along with these systems ensure our customers always get the safest, most quality-driven vehicles built.

Anthony Smith: Team Leader, Tool Crib

Add Value to the Organization by Developing Your People

Principle 9: Grow leaders who thoroughly understand the work, live the philosophy, and teach it to others.

- Grow leaders from within, rather than buying them from outside the organization.
- Do not view the leader's job as simply accomplishing tasks and having good people skills. Leaders must be role models of the company's philosophy and way of doing business.
- A good leader must understand the daily work in great detail so he or she can be the best teacher of your company's philosophy.

For this principle, I thought I should use an example. Charles "Chuk" Luttrell is our assistant general manager in the assembly department. He got his start in February 1988 on Chassis 1 in Assembly 1. He would start as a team leader at the age of twenty when our plant began. Chuk would get promoted to group leader in 1989 and would be sent to Japan for training on the brake and fuel systems. At the time it was located on Chassis 1. In 1997, Chuk would be promoted to the assistant manager of the trim section. He would become my first assistant manager.

In 1996 he would become one of the assistant managers of the in-house consulting group known as Operations Development Group (ODG) tasked with expanding deeper understanding of Toyota's production system through practical kaizen activities aimed at people development and overall KPI (Key Performance Indicators) improvement, especially in the areas of quality and cost. While in this job he would be very successful. One recommendation on Chuk's work is located on www.linkedin.com. It would be said that while Chuk was in this position, he would gain a broader perspective of all areas of TMMK. He quickly developed his capability in applying his knowledge on the shop floor. He also has a great sense to manage the interpersonal and environmental factors that impact the success of the initiative. As a result he was promoted to manager of this group while still participating in it. His leadership promoted the success of this group and TMMK.

In 1998 he would become the manager of Paint 2. This shop painted all the cars that went to Assembly 2. These cars were the first generation of the Sienna and the Camry. He would be in this position for three years and nine months eventually moving back to assembly.

In 2001 he would move back to assembly as a manager. He would be in this position through February 2007, when he would eventually be promoted to assistant general manager. This is the position he is currently in.

I had the opportunity to sit down with Chuk and ask a few questions for this story. I found this conversation to be very educational for me personally. I asked him what he had learned along the "Toyota Way." His first response was "Respect for humanity." When the Toyoda family built the first loom works and created jidoka, it didn't start with the idea of creating a great business. The idea was created because they saw they felt bad when someone would work really hard and then have to start over because the thread would break. So at the heart of Toyota is respect for humanity. The genesis of our company was based on this value.

I also asked what he would like to share about his career. I found this answer to be very compelling. To always have focus and persistence and that it is OK to make mistakes but to be reflective and always look to improve so the same mistake isn't made twice. It is also important to learn the Deming and Toyota principles and try to lead a business by following those ideas. That every day we should badge in knowing that we will do better today than yesterday. We should always look for kaizen and apply that to our personal performance and, finally, to always have the long view of any changes in mind. We need to think in terms of years before implementing any changes. For example, what will our cars today say about us in thirty years? Our legacy will always be that we are thinking forward. As a company, we always want to be ahead of our times.

When asked about the secret to his success, the answer was simple and straightforward: Even if we struggle and things don't go our way, we should never give up and we should not lose confidence in our abilities. We have to refuse to be or play the victim, and we need to focus on growing our circle of influence.

There is a sign hanging above our Chassis 1 line. This sign has been there since 1989. This sign has many names on it and is considered a sense of pride within our walls. On this sign Chuk signed his name as a team leader in 1988. I believe many team members and team leaders see this sign and realize that Toyota does believe in Principle 9. They really do believe in the idea that we promote from within. While we do hire from outside of the company to promote a diverse work culture, we do believe in the idea that TMMK workers can succeed and be more than just line workers. Chuk is a perfect example that hard work and determination can create a very successful career.

Principle 10: Develop exceptional people and teams who
 follow your company's philosophy.
   ■ Create a strong, stable culture in which company val-
     ues and beliefs are widely shared and lived out over a
     period of many years.
   ■ Train exceptional individuals and teams to work
     within the corporate philosophy to achieve excep-
     tional results. Work very hard to reinforce the cul-
     ture continually.
   ■ Use cross-functional teams to improve quality and pro-
     ductivity and enhance flow by solving difficult techni-
     cal problems. Empowerment occurs when people use
     the company's tools to improve the company.
   ■ Make an ongoing effort to teach individuals how
     to work together as teams toward common goals.
     Teamwork is something that has to be learned.

Jim Sivers: Team Member

We are glad that Toyota takes quality circles seriously. Our quality circle, Up All Night, in Assembly 1, second shift, Group IA-250 IP Line, is very active. A recent theme we chose to do dealt with safety. The following is a brief story about the theme with an interesting twist.

The IP Line is located at the south end of Assembly 1. We are the first production line every tour group sees coming out of the visitors' center. Our break area is directly across from our production line. We have to cross a major thoroughfare to get from our break area to our production line. The traffic around our break area and production line is constant. Tuggers from conveyance routes, vehicles pilot is working on, bicycles and team members walking to paint and body weld shops pass between our production line and break area. We remind each other constantly to be careful crossing the street.

Our quality circle noticed an abnormal situation. When production stops for breaks and lunch, team members seemed to let their safety awareness slip. As team members crossed from the production line to the break area, tugger and vehicle horns and bicycle bells sound off!! "Oops, sorry!" we would say. Not a good situation. Our quality circle felt we could fix this.

Putting the quality circle TBP system in action, our circle concluded that team members walked off the production line in any direction, not maintaining any awareness for safety. Our countermeasure for this was a gate and handrail system. This directed the exiting and entering of the team member to and from the production line in an orderly and safe manner. It made us pause and look before crossing to our break area. With everyone's agreement, our countermeasure was implemented.

We even presented this theme and were awarded the silver award.

Now the twist to the story: it was about two weeks after implementation of the countermeasures an accident was prevented. During production, a tugger pulling a full dolly of bumper reinforcements broke loose. The dolly was headed straight for one of our team members installing wires for the IPs at his process. The dolly crashed into the rail and gate system we had installed, preventing injury to the team member. Since this occurrence and implementation of our countermeasures, everyone maintains a high awareness of our exiting and entering our production line.

Once again, we are glad Toyota values the results from quality circles and their countermeasures. Thank you from the team members of the Up All Night quality circle: Deanna Mason, Scott Friend, Jason South, Jim Sivers, Bobby Parks, Lisa Noel, Michael Mullins, Lee Short, Chris Welch, Brian Vandrift, Dave Eades, Susan Menne, and Amy Tarter.

> To all the members of this quality circle, I thank you for your hard work and dedication. This kaizen truly kept a serious injury from occurring. Body weld first began using these gates a few years ago in their area with much success. This was the first example of protecting an assembly line with this kaizen. —Tim

Mike Bray: Supplier Perspective

In many ways, I guess you could say my association with Toyota has come full circle. Prior to coming to work for Toyota, I actually worked for a couple different companies that supplied the auto industry (Honda, Ford, and Chrysler, among others), but never had the pleasure of dealing with Toyota until I actually accepted that job as a team member on the assembly line. Having been in management at both companies, it was quite different being on that line, and while I have always strived to treat people right and do all I can to help others, that time spent on the line helped me to see clearly what makes Toyota so successful. The team members on the floor, the job they do, and a commitment to them from the organization they work for are simply amazing, and it cannot be understated just how important this is to Toyota's success.

Having worked for companies that were joint ventures with Japanese companies, I had a pretty good understanding for standardized work, kaizen, kanban, and so on before coming to Toyota, but had never seen it all put together in such a way as everyday life is at Toyota. Being a part of a company where it seems as though "good enough" is never quite good enough, and the fact that there is always room for improvement comes as an everyday way of life, has both challenged me and made me a much better manager and person. I am truly thankful for all of the opportunities working with Toyota has afforded me and how much it has taught me. Mostly, however, I am thankful for the wonderful people I have shared experiences with during my time there, and words can simply not explain enough how much that has and always will mean to me. I have some lifelong friends because of our journeys together, and I have enjoyed each moment together.

As I said, though, I have come full circle and now find myself as a supplier again. I am fortunate to work for a company that supplies Toyota, Honda, and Nissan. Believe it or not, after having stepped outside the walls of Toyota, it is amazing at how many simple pieces of the TPS puzzle that you just take for granted while you go about your everyday job. I work for a very good company, and yet as we implement things such as standardized work, kanban system, conveyance routes, and andon systems, it is amazing to stand back and see what a difference these things can make in a good, solid organization. I can only imagine how other companies struggle through times without these tools for success. Being a supplier to Toyota now, as I stated, I have seen firsthand their commitment to the supplier chain and that they fully realize that a strong supplier only makes them stronger. Toyota is totally committed to doing whatever necessary to assist their suppliers to become better and stronger.

While holding them to extremely high standards, they treat their suppliers with respect and make themselves available to teach TPS principles and strengthen their organization. They have supplier development groups that work with you to teach you the principles of TPS and problem solving, which only make every part of the chain stronger. My company recently had a product label issue, and this was the only issue of this type in nearly two years. Toyota immediately contacted us and set up meetings to come in and work with us on a countermeasure by sharing best practices that they had seen and implemented in other facilities. This signifies the commitment to excellence Toyota desires (only one issue in two years), but also the realization that we can always improve and their willingness to help us do so. By them coming in and allowing us to yokaten what they had implemented in other companies, it allowed us to evaluate our systems even further and implement countermeasures to make us even stronger.

Again, I cannot express enough how instrumental Toyota and its team members have been in my life, and I can speak from experience when I say that Toyota really does respect its network of suppliers by challenging and helping them to improve. Furthermore, I have the utmost respect for the team members who have left their mark on this book and the job they do every day, and am honored and privileged to be a part of it as well.

> Mike Bray is a highly admired former assistant manager from TMMK. He left Toyota for an opportunity to work for a supplier close to his family in Somerset, Kentucky. I am very thankful that he was willing to share some insight from his perspective of how Toyota supports our family of suppliers. —Tim

Continuously Solving Root Problems Drives Organizational Learning

Principle 12: Go and see for yourself to thoroughly understand the situation (genchi genbutsu).
- Solve problems and improve processes by going to the source and personally observing and verifying data rather than theorizing on the basis of what other people or the computer screen tell you.
- Think and speak based on personally verified data.
- Even high-level managers and executives should go and see things for themselves, so they will have more than a superficial understanding of the situation.

Philip Baugh III: Team Member, Mutilation Prevention

My journey at Toyota began August 18, 1997. I have worked on the chassis and door lines but the most rewarding has been in the areas of customer satisfaction, quality gates, and mutilation prevention. The team members at Toyota Motors Manufacturing, Kentucky, strive in every job to send the customer the very best vehicle that we have the ability to build.

We go to a lot of effort to create awareness in all areas of the assembly shop for mutilation prevention. One of the most effective ways of doing this was by taking the protective covers off of the cars and by having the team members treat the vehicles as though the paint were wet. This helped us create a no-touch policy, which is that if it is not totally necessary to touch the car as part of your job, we do not touch it. Also by empowering the team member on the process to be sure that there is nothing that could cause a mutilation, and if so to correct the situation as well as bring it to someone's attention.

Another way of dealing with mutilations is with a part of the problem-solving process or "plan, do, check, act" called genchi genbutsu.

Genchi genbutsu (stop and watch) has three parts: go and see, absorb, and take action. Even though we try to identify possible areas for a "mutilation" to occur, there are always more ways to continually improve.

We will keep striving for that to provide our customers with great quality.

> Philip takes his job seriously, and our cars are better because of it. He is the mutilation team member for Assembly 1. —Tim

Principle 13: Make decisions slowly by consensus, thoroughly considering all options; implement decisions rapidly (nemawashi).

■ Do not pick a single direction and go down that one path until you have thoroughly considered alternatives. When you have picked, move quickly and continuously down the path.

■ Nemawashi is the process of discussing problems and potential solutions with all of those affected, to collect their ideas and get agreement on a path forward. This consensus process, though time-consuming, helps broaden the search for solutions, and once a decision is made, the stage is set for rapid implementation.

"Take your time … Get input … Get it right."

Dave Garvin: Quality Circle Leader, Engine Line

People sometimes wonder why some things take so long to take place. Here at Toyota, it is no different. You must get input from everyone involved so as to make an informed decision. You need to think of the long-term effect. This brings me to a particular quality circle theme I would like to share with you to better help explain what it is I am trying to say.

This theme was on driveshaft scrap. Here at TMMK, we were having a lot of driveshafts that had to be scrapped in the plant. We started by looking at what exactly was it about the driveshafts that made it necessary to scrap them. We were experiencing dented dust rings, cut boots, and driveshafts that were being overextended. With the number of driveshafts that were getting scrapped in the plant on a daily basis, we had to act fast. But we couldn't just make a snap decision on a countermeasure; we had to think it through.

We had to look at several things that could possibly be causing the scrap. Was it the packaging, was it the parts handling, was it the way the parts were delivered line side, or could it be bad parts? We had to figure out where the damage was occurring in order to come up with a good countermeasure (a way to prevent the scrap). We had to get input from the team members from various areas involved so as to make an informed decision.

As you can imagine, this type of investigation takes time. When you come up with a countermeasure, you have to look at several things as well. You must first brainstorm ideas, then design something, then build it, and then you have to trial it to see if it is effective. Another thing to consider is the cost versus benefit. How long will it take to build? In this case with the driveshaft, whatever you come up with must be able to be modified in case of new driveshafts that may come along later with model changes. In addition to this, our circle would have to prove to the folks in Japan that we had developed a better way to handle the driveshafts, thus having to rewrite a technical information sheet (TIS); this outlines the way the driveshafts are to be handled. All of this takes time, and you must be patient as well as persistent.

You have probably heard the saying "Good things come to those who wait." I think that applies here. As part of our overall investigation on this theme, we determined that the line side and delivery racks were causing the in-plant scrap, and for one year alone here at TMMK, we were scrapping over $137,000 in driveshaft scrap. Due to our thorough investigation, getting input from everyone involved and taking the "Get it right" attitude, we reduced the driveshaft scrap to line side to zero.

This theme from start to finish, including the temporary countermeasure along with the permanent countermeasure, took approximately three years to complete. This is one of the extreme cases, But with a positive attitude, not giving up, and team members along with management here at TMMK as well as Japan working together, we were able to solve this issue, therefore providing our valued customers with a quality product. Not to mention all the people in all the different areas of Toyota we were able to meet along the way. In this fast-paced world, we must "Take our time ... Get input ... Get it right."

Principle 14: Become a learning organization through relentless reflection (hansei) and continuous improvement (kaizen).

■ Once you have established a stable process, use continuous improvement tools to determine the root cause of inefficiencies and apply effective countermeasures.

■ Design processes that require almost no inventory. This will make wasted time and resources visible for all to see. Once waste is exposed, have employees

use a continuous improvement process (kaizen) to
eliminate it.

■ Protect the organizational knowledge base by developing stable personnel, slow promotion, and very careful
succession systems.

How often do you second guess yourself? I know I do it all
the time. There are some times when I wish I could go back
and do things differently. This principle follows those same
ideas. Once we complete a project, we always go back to
evaluate how the project went and then put countermeasures
in place so the next time we do a project, we do better.

When all the other principles are followed, this principle
improves upon those. Standardized work can always be made
better. Our parts getting delivered to our processes can be
reduced to avoid having too many parts. We can reduce our
inventory in the field to make up for wasted ground to park
our cars on before they are sold.

We have a used car lot in Georgetown that always has very
nice cars. I look at these cars every day as I drive by the lot.
One time, I noticed a car that I really liked. Every day I would
look over to see if the car was still there. It sat there for a
couple weeks, and then I began to think that something must
be wrong with the car or it was priced too high. Sometimes,
having inventory too long will send out the wrong message.

Finally, we are always learning.

Chapter 13

Our Leadership

The greatest leader forgets himself
Good leaders support excellent workers
Great leaders support the bottom ten percent
Great leaders know that
The diamond in the rough
Is always found "in the rough."

—Lao Tze, 600 B.C.

"Thousands of minds, all working to 'do the right thing.'
Everyone understanding that 'we can always do better.'"

Raymond Bryant: Assistant General Manager, Assembly

I am an assistant general manager at TMMK in the assembly department. I started as a team member about twenty-one years ago, in the QC (quality control) department.

We often say that the most valuable resource we have is our workforce, the team members. I don't just believe this; I know it.

Our company functions on the basic principles of the Toyota production system (TPS). The Toyota Way guides us as we go about doing daily and longer term operations. When asked, "What is TPS?" the reference is usually to the tools widely talked about (andons, kanbans, etc.). As one of the many students of TPS (we all are still studying and learning), I have come to the conclusion that the success of TPS is in the commitment of our workforce to pursue it.

The tools are easily copied and put into place. They are, for all practical purposes, simple. The ability to understand and use them is equally as simple. The commitment to use them seems to be the struggling point for many organizations. Why? This is the "million-dollar question."

I believe that our success at TMMK is in the very simple understanding of what TPS really is. This I would define as:

1. Always doing the right thing.
2. Always knowing that you can do better.

The significance is in the "simplicity" of these two concepts. All of our workforce, each team member can understand, make decisions, and feel proud of these two guiding thoughts.

Pulling the andon is simply "doing the right thing." Alerting your staff or assistant manager when a problem has come up in a project is also done for the same basic reason. The one making the decision to notify and possibly seek help, along with the one that is receiving the news and/or request, both know that this is the "right thing" and expected in our culture.

This naturally carries beyond our work life into our personal life.

We sadly lost a team member some time ago in an automobile accident. Along with the strong support you might expect to see of the family during the days after this tragedy, many of the team members and leaders gathered together several weeks later to complete a home project that he had started. This was the "right thing to do."

My role is pretty basic in the grand scheme of things:

1. Setting clear goals that, if met, equal success (this requires face-to-face discussions to clarify and provide the why behind them); approving the ideas and methods suggested on how to reach those goals (this requires even more discussions, primarily listening, coaching, and the willingness to see many other points of view)
2. Supplying the support and/or resources needed (this really requires the highest amount of listening skill, because the secret is knowing when not to give input)

A successful manager knows not "when to say something," but when "not to." Thousands of minds, all working to "do the right thing." Everyone understanding that "we can always do better."

> Raymond Bryant, in my opinion, is one of the greatest servant leaders that I have personally ever seen. I appreciate him very much because I know he truly cares about his team members. —Tim

"Team members in every shop throughout Toyota think and act this way daily, and this is a clear reason that we are a highly regarded company and we will be here well into the future."

Kim Crumbie: Project Manager, Body Operations

I started my career with Toyota in Georgetown, Kentucky, in 1988 as a production team member in the body weld department. I have worked here for twenty-one years, and I am currently a project manager in body operations and have spent my entire time in body operations. I have worked in body weld production, stamping production, pilot (new model development), and production engineering.

When I was a youth, I always thought about "What will I be and what would I do when I grow up?" It was a scary thought that someday I would have to provide for myself without my parents' assistance. I was the type of child that was always inventing gadgets and loved the thought of one day building a robot, my own personal robot that would substitute for me at school so I could sleep in and one that would eat my dinner and look as if he enjoyed all of my peas and carrots while I watched from the closet. This dream of building my personal robot was something that I worked on for a while, and I made many attempts while I was growing up. Through trial and error, several slight electrocutions, and a small fire in my father's basement, I was successful at building a robot. It was sort of state of the art—by that, I mean I had to manually move the arms and hands while making my own sounds—but it was a robot nonetheless. I played with it all the time and never gave up on the dream that one day I would be working with robots.

That day was closer than I could have imagined as my father and I were watching the television and the local news reported that Toyota was building a huge automotive complex in Georgetown, Kentucky. At that time, it really didn't make that much of an impression on me as it did on my father, because I was in college and wasn't worried about anything. My father told me if I did not apply at Toyota, he would no longer continue to pay for me to keep going to Eastern Kentucky University. I decided it was in my best interest to go and fill out an application, not thinking that I would ever get a job. The testing was very extensive and thorough, and I did my best in each phase to complete the tasks. I didn't think I was going to be selected for the job because it had been a month since I heard anything from the recruiters. Then one afternoon, while I was at home, the phone rang and it was Toyota calling to offer me a job as a production team member in body weld. That day, my life changed forever.

Early on as team members in body weld, we were learning the Toyota Way. We immersed ourselves in the repetition of standard work. Our Japanese trainers coached us daily on the proper techniques and practices to be a good team member. At first, I really didn't clearly understand the reasons and rationale why we had to place the parts in the machines the same way every time, but the reasons soon became apparent. They were trying to teach us the value of standardization. It is the first step in the process and the reason for kaizen. Standardization requires everyone to complete the job the same way, every time; therefore, it becomes easy to find out why there is a problem or defect occurring. I began to understand the importance of standard work, jidoka, and just in time as a student in the welding department and how important it was for me to always improve my skill level.

Now multiply this logic times the thousands of employees at Toyota, and you can understand how this workforce joins together in a concept called "lean manufacturing" or the Toyota production system (TPS). TPS allows the proper actions to be coordinated in order to achieve organizational success.

This system empowers the team members to make decisions based on the needs of their area, and ensures the high-quality and safest process to produce the highest-quality vehicles possible. The culture inside the body weld is one of teamwork, efficiency, and attention to detail at all levels. It is the cornerstone of Toyota's success.

Since I started here as a team member, I have been given the freedom and ownership in the problem-solving process, which allows for practical, effective and smooth implementation of solutions to resolve issues. Allowing team members to help resolve issues is a key reason that Toyota has a clear advantage over its competitors, in my opinion.

My experiences are priceless, and I will never forget them. As a child I wanted to work with robots, and my dream was fulfilled as I was able to work in a department with multiple types of robots that lifted, welding, sprayed, checked, and performed many other operations. I have had the privilege to take over eighteen trips to Japan and visit many suppliers and give input on the next generation of equipment that was coming to our plant. There have been multiple visits in North America to vendors as well as meetings with designers to give feedback on new products. I feel blessed that I have had the opportunity to move to other areas within body operations for additional exciting challenges. Last year, as the production manager in stamping, my department received the prestigious Harbour Award for the most efficient stamping plant in North America. This was accomplished by the hard work of the team members that performed the daily tasks and drove the kaizen at their level to the support staffs like maintenance, tool and die, engineering, and a multitude of others that perform at this level every day, without the thought of reward but a sense of pride in what they do. Team members in every shop throughout Toyota think and act this way daily, and this is a clear reason that we are a highly regarded company and we will be here well into the future.

I've experienced many successes and challenges in my time with Toyota, but overall I have been pleased to be part of an organization that strives for excellence and provides a unique business model that has set itself apart from many of its competitors in the automotive industry. My experiences here will last a lifetime and have helped me to become a better leader, a better problem solver, a better employee, and a better person.

> I first heard Kim Crumbie when he was a guest on Fenella Smith's radio show. I was excited to ask him to participate after I heard his interview. Thank you, sir, for your participation. —Tim

"To fully understand this, you must put others' needs to improve the company first, and this will help strengthen what we call 'Team Toyota.'"

Dave Orrender: Assembly Manager

First, I must say, it is a privilege to work for the company TOYOTA. My career started back in 1987, and this, to me, is considered my "first real job."

I have been fortunate to have been promoted to the position of manager over my twenty-one years here at Toyota. As an employee, you go through several training modules (the Toyota production system, the Toyota Way, standard work, kaizen, just in time, as well as many others). All this training is a critical part of why Toyota is successful, but there is even a more important factor: the team members. You see, the knowledge of the system is needed, but those who execute it are essential. As a manager, it is engrained in you that the team members are what makes this company, Toyota, successful. We must gain their input, feedback, and buy-in because without each and everyone, we would not see our progress (kaizen) every day.

Working for and helping the team members are truly what make it worthwhile to come to the plant each day. As I have walked the production floors in assembly over these years, I have had many unique encounters with team members and how the Toyota production system is utilized.

I recall reviewing a team member's kaizen of making a job safer for his fellow team members. He and others in a group developed a way to improve installing the inner door trim pad. It was difficult to install, which created a safety concern, due to the door not being stable while pushing the trim pad to the door. They designed a way to keep the door steady while installing the inner door trim pad, which reduced the difficulty of pushing the part to the door. What is always good about kaizen is its ability to improve more than one aspect of the item you work on. In this instance, it improved the obvious, safety, but it also improved quality (the part fits better), productivity (no need to pull the andon to get support doing the process), and the morale of the team members (which is the biggest impact). What was most unique was not the fact the kaizen improved their process, but that someone took time to review and recognize them for what they had accomplished.

I could recite hundreds of examples like the one above. These have taught me how important one of the key phrases of the Toyota Way is, "Respect for humanity." To fully understand this, you must put others' needs to improve the company first, and this will help strengthen what we call "Team Toyota."

I'm glad to be a part of what happens in Georgetown, Kentucky, each and every day, and as I mentioned at the beginning of this story, this was my "first real job" and it still is.

Dave is my current manager. His patience with me and support for this book have been overwhelming. He truly does believe in and lives the phrase "Respect for humanity."

> Thank you, sir, for all you have done for me personally and for all the support you give to the team members on the floor. —Tim

"Toyota is always under strong scrutiny because they are considered a foreign company, but as a long-term member of this corporation I can honestly say they exemplify the values and business practices that all American companies should emulate."

Jeff Ayotte: Production Manager, General Assembly

Living and Learning the "Toyota Way"

I began my career at Toyota Motor Manufacturing in December of 1991 as a specialist in their manufacturing administrative office. This group had a multitude of assignments, but was primarily responsible for overall site management and coordinating activities for Mr. Tanaka, the senior executive coordinating officer of manufacturing (SECO). The SECO was a position held only by a senior Japanese executive who had many years of experience under the "Toyota Way" and was highly respected for his knowledge of the Toyota production system.

When interviewing with Toyota, I was interested in a position in engineering (since this was my educational background) but took this opportunity to become part of the Toyota family—a decision I have never regretted. The opportunity to work so closely with the SECO and other Toyota executives was an invaluable experience and one I will cherish forever. As a young man recently out of college, I learned Toyota's management principles and philosophies that built this company and made it one of the most respected in the world of manufacturing.

Toyota is unique in their management style and always strives to provide their employees learning opportunities. They encourage their staff to rotate positions and responsibilities in order to gain a broader perspective of the business and to become stronger individual performers. A traditional company usually has "specialists" where individuals become one dimensional and do not understand how their decisions can affect another area of the business. The Toyota approach, although risky, creates a stronger foundation where team members understand the business model and work together to achieve company goals.

This management style enabled me to take advantage of several opportunities and ultimately led me to my current position as a production manager in general assembly. I feel my career path has given me a strong foundation and understanding of the "Toyota Way," and has built my trust and admiration for Toyota as a whole. In my seventeen-year career, I have held positions in administration, engineering, line management, TPS advisor, and operations management. In each of these positions, I have had valuable experiences and thank Toyota for making these opportunities possible.

In addition to my career opportunities, I was fortunate enough to meet the love of my life at Toyota. My wife and I both work at Toyota in Georgetown, Kentucky, and met during a new model launch in which we were both assigned as leads in our respective areas. She was the lead administrative specialist for new model parts ordering (production control), and I was the lead assistant manager responsible for quality and process readiness within the assembly department. Although we very busy with our respective positions and responsibilities, we were involved in many meetings together where we got to know each other. One day I asked her if she would like to go out for dinner, and she agreed! The rest is history—she and I dated for one and a half years and then got married. I share this with you because if it not for Toyota and the opportunities provided to me, I would never have met this wonderful person who I can now call my wife. The blessings Toyota has bestowed on me go much deeper.

Several years ago, I went through a very difficult custody battle for my two boys from a previous marriage. If not for Toyota and their respect for humanity, I would not have been able to win this custody battle and have the opportunity to raise my boys. This was a very difficult process, but one my wife and I felt was necessary for the future of my young boys. Toyota and the management team I reported to were very supportive and gave me the necessary time to take care of this personal situation. The balance between work and personal life is something Toyota encourages, and the support I received enabled me to be successful at work and at home during a very trying time.

In addition to my boys, Toyota played a part in our other family blessing—the adoption of our little girl. The programs offered by Toyota enabled us to pursue a dream of adopting. This program is just one of many that are offered by the company, and another example of how Toyota is committed to providing that balance between work and family life. Our daughter not only became a part of our family, but a part of Toyota's family as well. She attended the Toyota day care since she was two months old, where she recently graduated "preschool," and is now doing extremely well in kindergarden. The development programs offered by the Toyota day care were very impressive, which prepared our daughter for the educational challenges she will be experiencing for the next several years.

In summary, my experiences at Toyota have shaped my life and made me a stronger person—one that is successful in both my personal life and career. Toyota is always under strong scrutiny because they are considered a foreign company, but as a long-term member of this corporation I can honestly say they exemplify the values and business practices that all American companies should emulate. The "Toyota

Ken Anderson: Manager, HR Safety

Finding our way ... the Toyota Way.

How did the sleepy little town of Georgetown, Kentucky, move into the global spotlight? What were the forces that generated such local pride and international partnerships? How did two countries with so much difference find so much in common? The answers to all these questions is incredibly complex, and yet painfully simple. All these things happened through "the Toyota Way"!

The Toyota way started for me as a summer college summer temporary worker in 1990. I had absolutely no idea what the inside of a factory looked like, let alone what it took to make a world-class vehicle. Like every other team member who walks in for the first time, I was blown away by the smooth "dance-like" motion that was performed every sixty seconds to produce a vehicle. The scope and complexity of Toyota were like a drug. And the further I looked into the Toyota world, the more I wanted to be a part of it. After six months, I asked (begged) to be hired full-time. My general manager did a very wise and noble thing, and I am still grateful to him today. He said that he would give me a full-time job with Toyota if I would promise to complete my education. I made and kept that promise.

I worked the engine assembly line and began to learn the Toyota production system. The Toyota culture was intoxicating. We were not just a team; we were a family. We were proud of who we were.

With the help of my team, I was promoted through the ranks and eventually to manager. Through the years, we have all studied TPS and tried to master the Toyota thinking way. In the beginning, we all questioned each aspect of TPS. Why should we run "just in time"? Why should we talk about things that failed and went wrong that day? Slowly we learned and began to understand how TPS made us stronger and more efficient. We began to ask "why" on our own. The Japanese taught us to ask "why" five times when we encountered problems. They taught us to value problems as lessons learned for the success of the future.

As a team, we began to be recognized for the quality of the product. And in time, we began to study and benchmark other automotive facilities. Through the years, many books and articles have been written about how Toyota encourages continuous improvement. In truth, our culture is very open to finding out which company is the benchmark, and then working to adopt their approach. All team members understand that we must never relax, and we must always give the customer more than they expect.

Other companies come to the Georgetown plant in hopes of learning the secret to our success. They usually take some ideas back to their home plant. But they never are able to put their finger on the magic of TMMK. Here's the secret. The magic is in our people. The team members at Toyota buy into the process. They work like they are stakeholders instead of employees. For this reason, and it is one very big reason, our secret is safe.

Over the years, we began to grow and expand our skills through TPS. We began to flex our kaizen thinking. We also began to focus on our safety. We learned that respect for people meant that we had to hold safety as one of our core values. We began to educate our team members as well as our management team. Our team members and our president began to ask better questions and seek long-term counter-measures for our safety concerns. We still have work to do to become the best in the world. But we know it is the Toyota Way. Just the fact that you are reading this book is proof that in the Georgetown Toyota plant, anything is possible.

> Thank you, Ken, for your support and contribution to the book. I feel that your story has tied every-thing together perfectly. You are correct in saying, "Anything is possible at TMMK," as long as we perse-vere and keep trying. —Tim

"Because of the great leader he is, Mr. Cho was promoted to chairman of Toyota Motor Corporation in 2006."

Mr. Fujio Cho: Chairman, Toyota Motor Corporation

Nila Wells: Specialist, Community Services, TEMA

I am honored that I was asked to write about one of the most influential people I have ever known. How do I begin to describe a person who encompasses every aspect of what it means to be a great leader? I will do my best to give justice to my friend and mentor, Mr. Fujio Cho.

I was hired in February 1987 by Alex Warren, then the vice president of administration, to assist him; our executive coordinator, Hiro Adachi; and TMM's executive vice president, Fujio Cho. I met Mr. Cho for the first time in April of that year and began an experience of mutual respect and trust that will long live in my memory.

What were the qualities that Toyota Motor Corporation wanted when sending someone to Kentucky to oversee its first wholly owned plant in America? My thoughts are someone who is well respected, someone who has strong integrity and is a team player, and, of course, someone who would be able to teach the Toyota production system to team members with little or no automotive experience and have them build a quality vehicle. Mr. Cho certainly was the appropriate choice for all of these reasons and so much more. So, my thoughts may not be in the order of importance, but certainly in the order I think when I recall my time with Mr. Cho.

When I think of Mr. Cho and his leadership, it is hard to imagine that anyone would be able to emulate the qualities that make him the person he is and, in turn, a great leader. What comes to mind first for me is his infectious smile. If he was having a bad day or not feeling well, you never knew because he always had a smile for everyone. When I think of his smile, I can't help but smile, too.

Mr. Cho's true concern and respect for team members spoke volumes about the type of person he is. I remember Mr. Cho always taking the time to walk the plant floor, talk to the team members, ask how they were doing, what problems they were experiencing, and taking the time to listen to how they thought the problems could be best solved. But it didn't stop there. He would make sure the problems were corrected and the team members' concerns were addressed. Respect for people just came natural to Mr. Cho. And, all who knew him gave him respect in return—with him, it would be impossible not to!

It would seem to me to be very difficult to be a leader and not show anger and frustration at times. In all the years I have known and/or worked with Mr. Cho, I have never seen or heard him speak disrespectfully to anyone or of anyone. It is a very rare and special person who never speaks in anger or raises their voice to motivate people to get the job done and done well.

I remember the pride Mr. Cho had when the first Camry rolled off the line in May 1988. You could see in his face that he was so appreciative and proud of all the team members and their hard work.

Mr. Cho's sense of community is another of his great leadership qualities. It was very important to him to live in Georgetown when he moved from Japan. He wanted to be a part of the community and know his neighbors and let them get to know him. This resulted in a mutual trust and respect not only for him personally, but for Toyota in the community. People in Scott County speak very fondly of Mr. Cho and will tell stories of seeing him take a stroll down Main Street or fishing on the Elkhorn, or how they were invited into his home for dinner.

Mr. Cho's home was not only open to the community, but also to team members. On many occasions, he would invite groups into his home to enjoy karaoke. One of my fondest memories is when Mr. Cho had a dinner party in my honor in appreciation of the work I did for him. He invited my husband, coworkers, and their spouses. He even assisted his wife, Emiko, in preparing the meal. This was to be just one of the many ways he showed his appreciation to me.

So, it's very difficult to identify just one quality that makes a great leader, especially when it comes to leaders like Mr. Cho. There are so many that make the whole, and Mr. Cho is the one person I have known that encompasses them all.

TMMK was very fortunate its first leader was Mr. Cho, who created a firm foundation for us in the United States as an automotive-manufacturing facility. Because of the great leader he is, Mr. Cho was promoted to chairman of Toyota Motor Corporation in 2006.

> One of my goals for the book was to have someone write a story about Mr. Cho. As I was getting closer to completing the book, I found out that Nila worked with him. I appreciate her willingness to share this part of her life with us. —Tim

Gary Convis: President, 2002–2005; TMC Managing Officer, 2004–2007; CEO, 2005–2007

Jacky Ammerman, CAP: Associate, Toyota North American Production Support Center, Georgetown, Kentucky

As I participated in what I now refer to as the "best interview of my life," I came to realize that the stranger from California had a true respect for honest, down-to-earth individuals. He was interested in my thoughts on business matters. He understood that I felt the interview was as much about me interviewing him as it was about him interviewing me. And he appreciated the fact that I am not a "yes" person. Little did I know at the time what a life-altering event this interview would be for me.

I was expecting a phone call from HR to receive a status update regarding the open position ... imagine my amazement when he extended the personal invitation to be a part of his team! That's right ... Mr. Convis walked up to my desk, shook my hand, and asked me if I would please be his assistant!

Mr. Convis, who I now refer to as Gary, was the first American to be appointed to the position of president of a Toyota facility. This also meant that he was president of the largest Toyota facility outside of Japan. Even though this man was taking on such a monumental task, he never lost sight of the importance of respect for others. He once said, "We have to draw upon the experience and expertise of our Kentucky team members." This was not just an offhand remark; Gary truly believed this—he always believed strongly in the Kentucky workforce.

It was truly a surreal experience for me ... a southern gal raised in Kentucky with a deep southern accent was the assistant to an international leader in the automotive industry. Gary made me feel so comfortable that many times I had to remind myself of the importance and level of his role in both North America and global automotive industry.

I feel that we made an outstanding team—there was complete trust between the two of us. Gary always welcomed the sharing of thoughts and ideas—a person's position in the company had no relevance. I watched in awe as Gary worked passionately to mentor leaders for the Toyota of the future. Gary loved to impart his wisdom on others. It was amazing how knowledgeable he was. Even though he came to TMMK with a monumental task of teaching us how to tighten our financial belt, he aspired to make us, the TMMK family, the best that we could possibly be.

During his tenure, Gary did receive some less than favorable feedback from many team members due to some of the financial cuts that he initiated to secure our business condition. This is a great example of Gary's visionary ability—he felt strongly that we should strive to save funds where there was an opportunity. And today, TMMK remains one of Toyota's most financially sound facilities. While we have experienced reduced production numbers during this recession period, we have been able to continue the cost-cutting practices, keeping our doors open and our team members working.

In addition to being such a visionary leader, Gary is also a devoted husband, father, and grandfather. One of the biggest challenges that he faced each day was the balance between his work and personal life. Many people were not aware that Gary had a special needs son. It warmed my heart to hear about their times together when Gary would take Kevin out to a movie and/or out to eat. And when Gary would receive a new picture of one of the kids or grandkids, his face beamed with pride. Again, it is hard to think about men of this level of leadership as being "regular family guys."

Now retired from Toyota, Gary continues to maintain his longtime relationships and strong bonds with his Toyota family. I feel very fortunate to count Gary as a friend and mentor in my life. I learned so much as I worked "with" him. I never worked "for" him—Gary always treated me with great respect and looked at me as an integral part of his team.

Steve St. Angelo: President, TMMK

Rebecca Lucas: Associate to the Chief Quality Officer,
Toyota North America

As an associate working in the chief quality officer and former TMMK president's office at the largest manufacturing facility in North America, I was honored when I was asked to provide some insight in my role of supporting the president. How can I tell you about Steve St. Angelo? Perhaps a story or two will help you understand why I enjoy working for him.

Importance of a Caller ...

The cell phone buzzed ... people's heads turned as the young girl whispered to the caller, "I am in church and will return your call when I'm out." She hadn't recognized the number and didn't know who the caller was. After the Sunday services concluded, the girl kept her word. When the call was answered, she said, "I am returning your call—may I ask who this is?" The voice on the other end said, "This is Steve St. Angelo, president of TMMK." The girl had called his office earlier in the week to request that he reconsider canceling the annual TMMK company picnic held at Kings Island amusement park. The recent economic downturn made it necessary to cut activities, and the family picnic was suspended. However, after many negotiations, the company was able to offer discounted tickets to the amusement park so team members and their families could still continue their annual tradition.

This is only one of many situations that illustrate Steve's willingness to listen, whether to a team member or family member. He really listens to people and offers his best advice.

The Expired Passport ...

It was Monday, Memorial Day, and I was enjoying the last day of my extended holiday weekend when my cell phone rang. It was Steve on the other end calling from the Detroit airport to let me know that his passport had expired in early May.

Now, knowing that he had to be in Japan for a presentation on Wednesday, I knew that time was of essence. I quickly thought to myself, "Where does a person possibly begin to get a passport renewed overnight?" and as if that wasn't bad enough, on a holiday to boot. I spent hours on my home phone, cell phone, Blackberry, and computer making the necessary contacts in hopes of obtaining a renewed passport before his flight departed. It was down to the wire.

Steve had to stay overnight in Chicago and then walk around the corner from his hotel (or so we were told), which in reality was a three-block walk in 90 degree heat and humidity levels exceeding the temperature.

After arriving at the passport agency and waiting in a long line, the clerk told him to come back at 1:00 p.m. for his newly issued passport—the exact time of his plane's departure. Steve explained that it was imperative that he get his passport much sooner. He returned to the passport office at 11:00 a.m. to find that his renewal had been expedited and was ready. He rushed out the door, flagged down the first taxi, and was off to the airport to catch his flight to Japan.

Many would have been angry and frustrated, but Steve's calm personality and demeanor let him take it all in stride. Instead of being upset, he was very grateful and took the time to personally thank me for my support in helping him make the trip to Japan. His humor radiates whenever he retells this story.

Child Development Center ...

Every day is a new adventure in the office of the president, whether it's a normal day or a holiday. Speaking of holidays,

Steve is passionate about dedicating time to various community groups as well as the many TMMK activities. One of his passions is his annual visit to our Child Development Center dressed up in a festive costume, distributing gifts and candy to the team members' children (not just during the holiday season, but throughout the year). We have video of Steve bursting through the classroom doors to loud welcoming shouts of "Steve!" as the children race to give out hugs and collect their holiday gift and candy from Steve's holiday decorated tote.

When I think of the Child Development Center, one story comes to mind. After one of Steve's visits, I received a call from a parent whose son was at the facility. She wanted to personally thank Steve for the kindness and generosity he showed the children during a recent holiday visit. The parent said that her son was filled with joy and excitement, all from just a simple gesture of kindness from Steve, stating, "This is what being a part of the Toyota family is all about."

A Man of Many Hats ...

Steve is much more than just fun and games, though. You may have seen him in the plant donning a red felt holiday fedora. That's kind of symbolic of the "hats" he wears every day: chief quality officer for Toyota North America; executive vice president of Toyota Motor Engineering Manufacturing North America; senior vice president of Toyota North America, Inc.; managing officer for Toyota Motor Corporation; and former president of TMMK. Many don't realize how many different hats Steve wears each and every day. He is not your typical pampered, fawned-over executive. He loves his family, his team, and his work.

Steve is very adamant about the "open-door" policy that he implemented when he arrived at TMMK in 2005. There are days he is inundated with calls from customers, vendors, team members, and others—on top of his eight-hour day being overbooked to ten hours and twelve hours on a regular basis.

Somehow he always finds the time to listen, return calls, or personally walk through each manufacturing area checking on team members and their well-being.

Steve's job responsibilities don't afford him normal day-shift work hours, as he frequently has meetings, teleconferences, and video conferences until 11:00 p.m. He even occasionally has 2:30 a.m. calls, yet still arrives at work the following day on time and with a boisterous voice welcoming "the team." Even though some decisions are not popular with all the team members, Steve knows his decisions have to be made for the success of the entire team's long-term vision.

The Effects of the Economy ...

The 2008–2009 economic downturn has been a huge challenge for everyone. One team member told me of driving through Georgetown trying to imagine it looking like Flint, Michigan.

The rumor mills here at TMMK were the only thing working overtime when one meeting was announced—a meeting about which there were whispers of the National Guard attending.

Everyone had their best poker face firmly set when they came into "The Meeting," and all were braced for the worst. Then, the news was revealed—Toyota executives would be taking a cut in pay instead of sacrificing our jobs. In Steve's case, that meant a loss of more than a third of his salary—but you'd never know that from looking at his face.

In reality, there was never a threat to "our jobs." Some things simply were not on the table for negotiations.

Not many days later, we saw the heads of other automakers arrive in Washington, D.C., in their private jets. In contrast, at Toyota we followed our principles and "protected our castle," with Steve leading the way.

Gone are the days of executive privilege that Steve enjoyed at his previous employer, where he was employed for thirty years. He had such amenities as his own parking space near the front door and a private elevator to his office with a great view, where he was greeted each day with a Danish and coffee. Steve claims he doesn't miss the perks.

Steve St. Angelo joined TMMK in April 2005. Prior to his arrival, I did not know what to expect when this former executive from another company was to become my new boss. However, much to my surprise, he was not the typical "suit" that I had imagined. He brought a wealth of knowledge and experience to Toyota that he had gained throughout his career. We were very fortunate to have him here to lead our team during this recession.

So, let me leave you with one more story about Steve.

During Steve's first week at TMMK, we had the following exchange. "Rebecca," he asked, "do you think that facilities knows I work here?" I looked up and replied, "Yes, I'm sure they do; is there something wrong?" Steve said, "Well my trash is still here," to which I replied, "We each have to empty our own garbage and segregate at the environmental recycling center." I turned back to my computer, when I heard him say, "Rebecca, after lunch, will you show me how to dispose of my garbage?" I did, and he faithfully recycles at the end of each day just like the rest of us.

Personally Shakes the Hands of 7,000+ Team Members ...

I have worked with Steve for four years, and I can certainly tell you that he is much deeper and personable than most people imagine. He is a very genuine, compassionate, and caring individual. It still amazes me that he would take time from his hectic day to stand at the entrances across the plant and give out ice cream to team members as they were departing or arriving for work. He did this just to show his appreciation for their hard work and dedication to TMMK. Each year during the holiday season, Steve sets aside one week to personally walk through the over 7.5 million square feet facility and shake the hand of every team member. He does this simply to express his appreciation and gratitude for the hard work and dedication that each team member brings to the success of our facility in Georgetown, Kentucky. He has the utmost respect for every person, regardless of race, origin, title, or anything else.

Genuine and Compassionate Person ...

I have enjoyed sharing a few of my stories about Steve and demonstrating to you the kind of person that he is. It is truly my pleasure to have the opportunity to work with such a genuine and compassionate person. His sincere and caring nature is too rarely seen in the business world, especially at the executive level. He is always willing to go out of his way to ensure that team members' issues are addressed. That's why I am telling you that there is a reason that he is our leader.

> Rebecca has become a great friend. She has helped me in many ways through this book and other projects that I have worked on. She is extremely busy but always takes the time, and for that I am very grateful. —Tim

Chapter 14

Our Ambassadors

Our company has faced a lot of trials this year. We had a lot of people in the media be quick to jump to conclusions and make assumptions. We had UAW people picketing Toyota dealerships around the country trying to get us to go union. They made one big mistake by doing that. The union bosses proved why we don't want or need them. If they are willing to threaten our customers and keep us from selling cars now, that proves they don't care about us or our well-being. All they want is our union dues, and *we* realize that.

Through those trials, we had many people come out and defend us. This chapter gives just a few examples. We had everyone from the governor of Kentucky to a nineteen-year-old college student in Florida "fighting for us." Each person was needed and appreciated for standing up for the 37,000 Americans that work for our company.

Georgetown Mayor Karen Tingle Sames

In 1986, I thought that the most important thing in my life was the fact that I had graduated from Georgetown College. But in 1986, something bigger was happening and I really didn't realize what an impact it would make on my life. The same week that I graduated from Georgetown College, Toyota Motor Manufacturing came to the city of Georgetown and had a groundbreaking ceremony. I was too involved at that time with myself and my own plans and aspirations to even think about what an impact this would make on my life and my future. Toyota has made the largest impact for our city, our community, and even our state, more than anyone could have imagined in this area.

I grew up in Georgetown, went to school here, and am the fourth generation of a family that lived in Georgetown. In 1986, our whole world changed. Some that were here in this city felt that Toyota was a negative, but over the past twenty years Toyota has been the most positive influence in all of our lives who live in this community. Opportunities have been presented to each of us because of Toyota being here. When Toyota was announced, the population of this city was less than 10,000 … now we are over 24,000 people, making us the seventeenth largest city in the state of Kentucky. Up until that time, very little focus was about what was happening in the city of Georgetown. Now the world looks to us; international culture has been brought to Georgetown due to the influence Toyota has brought.

Toyota has been a strong partner with the city of Georgetown and the community. They came here with the intent to be a good partner, and they have been. Georgetown has changed since 1986. In 1986, the Scott County government was more important than city government, but over the last twenty-three years, city government has grown tremendously and has become the driving force of the community. During the time, Toyota has demonstrated what teamwork and partnership are all about. Not only do they teach that to their team members, but it is taught throughout the community. They are here to lend a helping hand and lend assistance at any time, and they do.

The helping hand of Toyota has allowed us opportunities for our city in ways we would have never have thought possible. We have the best parks and recreation program, second to none in the state. We have an airport that accommodates international guests. Our city government departments have grown over the years and have become stronger to offer the best possible service to our community. I would never have believed that I would grow up to be able to be mayor of this city and watch how it has quickly developed over the twenty-three years that Toyota has been here. I am proud to be a part of this city, and proud to have Toyota as our neighbor in partnerships.

With Toyota coming to our community, we have been able to keep our traditional values, been able to keep our southern hospitality, our southern charm, and our delightful downtown. And we still celebrate the talents of our people. The people that Toyota has brought to our community have strengthened the community and made it even greater. So I want to take the opportunity to thank Toyota for its vision to come to Kentucky and for all the people that made that happen.

> Mayor Karen Tingle Sames is a wonderful person and a great leader in our community. Before becoming mayor, she worked and helped manage her parents' flower shop called "The Carriage House." Our mayor did the flowers at my wedding in 1999. Even though I can't remember myself, to this day I could call Karen and ask her to send my wife a dozen roses for our anniversary, and they are always the flowers that we had at our wedding. (Of course, I quit doing this once she became our mayor.) In my opinion, a person that can remember such small details as this fits the mold of a wonderful leader. I also go to church with Karen and her husband Kevin, and consider them to be great friends. —Tim

"I Support TOYOTA Workers—They Are Still the BEST!!!!!"

Leigh Ann Reeves

I am the proud wife of Rob Reeves, a specialist in quality engineering at TMMK. Our family's journey with Toyota began in 1988, when we were a young married couple with a newborn baby. Rob had graduated from the University of Kentucky in 1986 and was employed as an operations manager at a local communications contractor. His title sounded impressive, but the pay was barely enough to support our little family, especially since approximately a quarter of his pay went toward our health insurance each month. To help make ends meet, I was working in the parts department at a major retailer when a coworker of mine began to tell me how her neighbor had just gone to work at the new Toyota Plant in Georgetown and that he was making a very good salary. This got me to thinking that TMMK might be a good place for Rob to work, so I asked him to update his résumé and give me several copies. At that time, there were Toyota job postings in the *Lexington Herald Leader* every Sunday, so I started watching them and it wasn't very long before I saw one that I knew Rob was more than qualified for. I stuck his résumé in an envelope and put it on our mailbox for the mailman to pick up the next day. The funny thing was I didn't have any stamps, so I taped twenty-five cents to the envelope hoping that the mailman would take it, and, bless his heart, he did. Shortly thereafter, Rob had an interview with Mike DaPrile, who was the assembly plant manager at the time, and within a few weeks began his career with Toyota as an assistant staff in assembly production engineering. What a wonderful blessing from God this was for our little family.

Over this past twenty-two years, I have been blessed to have the freedom to be a stay-at-home mom, to be there for my extended family during critical illnesses, to go back to school and take business classes, and, most recently, to start a new business. We're blessed, as a family, to be on our second home, to have owned and driven a number of quality dependable Toyota vehicles and take some wonderful vacations. We have enjoyed many company-sponsored family fun activities such as the annual company picnic and Easter egg hunts, and have appreciated the opportunity to purchase discount tickets to the zoo, amusement parks, concerts, and other special events. Rob and our son Daniel are very involved in scouting, and Toyota has allowed Rob to volunteer with the troop and to earn money for the troop through the company's Volunteers in Place Program. I can't think of too many companies who take care of their employees and in so many ways like Toyota does.

Toyota has not only supported our family monetarily but emotionally as well. Approximately two months after Rob began working at TMMK my father was suddenly and tragically taken from us, and it was a huge relief for Rob to be off, with pay, to be with me during that time, and I couldn't believe the big, beautiful flower arrangement they sent to the funeral home. A few years back, when my stepfather passed away, a representative from Outreach came to the funeral home and they sent flowers and a food basket as well. They celebrated along with us when our son was born by sending flowers and a personalized baby book. Just this past summer, Rob came down with a tick-borne illness that landed him in the hospital for a week. TMMK Outreach was on the phone with us the day after he was admitted, asking what they could do, and they quickly brought me food vouchers that the kids and I were able to use in the hospital cafeteria. As we all know, these unexpected life events can really blow your household budget, so it was really nice not to have to worry about our meals that week. These are just a few examples of the many ways that TMMK has supported and continues to support our family and why I am its biggest fan.

Rob has been in quality engineering for twenty-one years now and takes his job very seriously. He knows that people who buy Toyota vehicles expect a quality-built, safe, and reliable vehicle, and he does not want to let the customer down. He has worked long hours and traveled to Japan, Mexico, Canada, and all over the United States, sacrificing many family hours to ensure public safety. As much as I miss him at times, I am also extremely proud of the work he does and am happy to support him so he can be at his best for the company. Through the years, I have come to realize that by supporting him in his job, I have become a TMMK team member as well.

In January of 2010 when the news stories, alleging sudden unintended acceleration, began to break and Toyota began to issue recalls and stop production, I was concerned but I had the utmost confidence that the company was investigating the issues and would provide any needed fixes in a timely manner. In fact, my pride in the company did not waver one bit. So you can imagine how personally I took it when I heard the anchor of a major nightly news program reporting on the "black eye" on Toyota vehicle quality. This, along with other relentless national news reports and our own local stations' negative spin on Toyota recall stories, got me to thinking about what I could do to help my husband's faithful employer. I knew that the local stations had to report the recall stories, but I thought they owed it to TMMK to also report on all the good ways Toyota has impacted our area in the last twenty-three years. Some of my story ideas were Toyota's positive economic impact on central Kentucky, how Toyota encourages its employees to volunteer their time with charitable and civic organizations, and the history of Toyota in central Kentucky. Being a self-proclaimed Facebook addict, the idea hit me one evening to start a Facebook group in support of TMMK workers. My goal was to invite all my friends to join the group and ask them to invite their friends and try to get to 1,000 group members. Once I got to 1,000 group members, I felt that I would have some power behind me to ask the local media to balance their reporting with positive Toyota stories. So on February 6, 2010, the "I Support TMMK Workers—They Are Still the BEST!!!!" group was born and a new adventure began.

It wasn't even a week before I had met my 1,000-member goal. On one hand, I was shocked; on the other hand, I knew there were a lot of good people in central Kentucky who were gladly standing behind Toyota during their biggest crisis ever. I began posting the few positive stories I had heard reported though various media outlets, encouraging our group members to invite their friends, and I contacted the local media as I had originally planned. I was interviewed by all our local stations, one local newspaper, and even a Japanese film crew. It wasn't long before word of our group spread beyond central Kentucky, and team members from Toyota plants in the U.S., Canada, and around the world began to join our group as well. Of course, at this point it made sense to change the group name to "I Support Toyota Workers—They Are Still the BEST!!!!"

During this time, it was announced that House Oversight and Government Reform Committees of Congress were going to hold hearings to look into the Toyota recalls. I began encouraging our group members to show their support in any way they could. One simple idea I had was for people to use those auto glass markers to write "I/We Support Toyota" on the back glass of their vehicles. Several people jumped onboard with that idea, but a lady named Lisa Doan had a better idea. Lisa contacted me and told me that her husband Brad, a second-shift team member, ran a small sign business in his spare time and he could custom-make vinyl window letters that said "I/We Support Toyota." Brad began producing all the vinyl lettering he could in his spare time, and we began distributing them all over central Kentucky and mailing them all over the United States. Another team member, DeLloyd Smith, who had a screen-printing business, began making "I Support Toyota" T-shirts for us as well. It was cool to see the TMMK workers and workers at other plants around the U.S. sign banners, which were then delivered to service departments around the country thanking them for taking care of Toyota customers during the recalls. Customers began delivering baked goods and treats to their service departments, showing their support and appreciation. I have heard it said that tough times have a way of bringing out the best in people, and I think that Toyota and its team members, dealers, technicians, and their families and customers have proved this statement true once again. Currently, we have approximately 12,300 members of our Facebook group, and the investigations have found no problems in the electronics of Toyota automobiles.

Thousands of recalled vehicles have been repaired, and Toyota is number one in retail sales. It has been a tough year, but it has been a great year. Through all of this, we've worried about the future of the company but have found out just how strong it really is. The Bible says to give thanks in all things, and I am giving thanks for what we have all been through because we are all better for it and I have never been prouder to say, "I Support TOYOTA Workers—They Are Still the BEST!!!!!"

> Leign Ann: I believe that out of every struggle comes a silver lining, and getting to know the Reeves family through all of this has been a true blessing. —Tim

State Senator Damon Thayer (R-Georgetown)

Since its decision to locate in Georgetown was made twenty-five years ago, Toyota has been a model employer and an ideal community partner for my hometown. While other businesses ship jobs overseas, Toyota has consistently upgraded and expanded its facilities because of the fine work Kentucky workers have done. Ninety suppliers have located here in Kentucky, with more than 10,000 jobs benefiting our economy.

The Toyota plant in Georgetown, like seven others nationwide, recycles its waste to cut down on landfill use, and the on-site vegetable garden produces more than 1,000 tons of food for the local food bank. In all, more than $37 million in donations and contributions to local charitable groups have benefited our community.

The company's innovations have revolutionized the industry. When people think about a hybrid car, an image of a Prius naturally pops into their minds. Toyota's just-in-time supply chain management philosophy has saved businesses in all sectors millions of dollars—savings that have been passed on to consumers.

In light of all these facts, I can't help but be immensely proud to have Toyota in my community. The incentives we offered to Toyota were the smartest investment Kentucky ever made—and perhaps one of the smartest public investments in U.S. history.

That's why I was so shocked when Toyota was brutally attacked by our federal government during its recall efforts. The same national leaders who bailed out big banks and billionaire Wall Street tycoons were quick to denounce Toyota, a company that, with its suppliers, employs more than 200,000 people throughout Middle America. At a time when consumer spending was desperately needed to stimulate the economy, President Obama's secretary of transportation told people not only to stop buying Toyotas, but not to drive the ones they already owned! In short, many public officials made hasty assumptions without considering the consequences of their actions or waiting for all the facts.

These were attacks not only on company leadership, but also on the men and women who had spent long hours in the factories building the bestselling cars in the nation. They were attacking the people of my community, and I couldn't sit still. I wrote op-eds defending Toyota and sent them to any outlets willing to tell the truth. I sponsored legislation honoring Toyota, legislation that passed the Kentucky State Senate, so people would understand the value it brings to our state.

True friends stick together through thick and thin, and Kentucky has had no better friend than Toyota for the last quarter-century. True to form, Toyota moved quickly to make sure its vehicles were safe, auditing their entire manufacturing process, working with suppliers to improve quality, and training workers to make sure Americans can count on their products.

In hindsight, I'm sure the alarmists regret their words and actions. I've never regretted mine. The partnership between my community and Toyota has paid incredible dividends, and will continue to do so long after irresponsible public officials have been driven from office.

> Senator Thayer, thank you so much for writing and continuing to support our company and being such a strong leader in the state capital. The people of Kentucky are grateful that you serve. —Tim

Curtis Holmes: Loyal Toyota Customer

Do you feel that there are safer vehicles on the market and roads besides Toyotas? If so, does the automotive manufacturer you prefer make all of its security features standard on all makes and models, or does it nickel and dime you for each individual feature? If you go to the dealership and find out that you need to upgrade the vehicle package, therefore increasing the price of the vehicle, in order to get more safety features, then it sounds to me like the company puts money first, not your safety. Now tell me, does that sound like a company who's concerned about its customers' safety and assurance, or is it self-centered and only cares about the almighty dollar? Is that really a company you want to entrust your family's life to? Toyota does not do that. Toyota does not offer a vehicle with little or no safety features. You cannot buy a Toyota vehicle that does not have every available safety feature on it. Every single vehicle make and model produced by Toyota, Scion, and Lexus comes standard with thirty safety features which no other manufacturer offers, and they do that without increasing the price of the vehicle to you, the customer. I personally believe there is no safer vehicle than a Toyota. Its "Star Safety System" includes an exuberant amount of safety features, which no other car manufacturer offers. Toyotas are extremely safe vehicles with state-of-the-art technology and will protect you and your loved ones at all costs.

The Star Safety System includes a large number of security features for the exterior of the vehicles. All Toyota vehicles include features such as vehicle stability control (VSC), which, through sensors, prevents wheel slip and loss of traction by reducing engine power and applying brake force to the wheels which are slipping in order to keep the vehicle straight. Another one of Toyota's safety features is a traction control system (TCS), which helps maintain traction on wet, icy, snow-covered, loose, and uneven surfaces by applying brake force to the spinning wheel(s), which has (have) lost traction, even to the nonpowered wheels, such as the rear wheels on a front-wheel-drive vehicle. A third safety feature Toyota has standard on all vehicles is antilock brakes (ABS), which prevent the wheels from locking up when the brakes are aggressively applied to avoid a collision. This is achieved through pulsating brake pressure sent to each wheel independently to maintain control in emergency braking situations, therefore avoiding a skid. A fourth safety feature is electronic brake-force distribution (EBD), which helps keep the vehicle more stable and balanced when braking. Integrated sensors in the vehicles sense which part of the vehicle contains the most weight, for example the trunk, due to cargo, and apply more brake pressure to the wheels supporting the most weight and decrease the breaking pressure on the wheels with the least amount of weight on them, which greatly improves braking efficiency.

Every Toyota vehicle is also equipped with smart stop technology (SST), which automatically cuts engine power when both pedals are pressed at the same time, causing the vehicle to come to a nice and safe complete stop. All Toyotas come standard with a direct tire pressure monitoring system (TPMS), which constantly monitors the air level in each individual tire and alerts the driver when the air pressure in one or more tire(s) is above or below the manufacturer's recommended tire pressure. Having underinflated or overinflated tires not only causes uneven wear and tear on the tires, but it also greatly affects the vehicle's braking distance as well.

Toyotas also have front and rear energy-absorbing crumple zones and side impact door beams, to protect the occupants during a collision. Another safety feature Toyotas have is hill start assist (HSA), which prevents the vehicle from rolling backward while taking off from a stop on an incline. A twelfth security feature is a collapsible steering column, which means that upon impact of the steering column, the steering column will collapse, preventing the steering wheel from being pushed into the driver's body. All Toyota vehicles also have an engine immobilizer system (EIS), which means that it is impossible to hotwire the vehicle or have it stolen. The vehicle will not start unless the correct key fob with the correct ID is inserted into the key-fob slot or sensed via the smart key system. There is also a center high-mount stop lamp (CHMSL) on Toyota vehicles, which makes it easier for drivers of larger and taller vehicles, like the Toyota Tundra or Sequoia, to see the brake lights and prevent a rear-end collision. Toyotas also include a temporary spare tire.

Toyota vehicles also include many interior safety features, such as an advanced airbag system (AAS), which means there are sensors in each seat which weigh each occupant and automatically program each individual air bag to deploy at a different rate of speed and pressure to increase the safety of each occupant and lower the chance of injury due to impact from the air bag. It can even automatically turn off an individual air bag if the occupant is of a very low weight range, such as a toddler in a car seat. There are ten air bags which accompany this system. There are four curtain air bags, two front-mounted upper-body air bags, two front-mounted knee air bags, and two seat-mounted hip air bags. Each individual air bag's force and speed are automatically adjusted depending on the weight of the passenger in the seat. Toyotas also include active headrests, which as soon as the air bags deploy, automatically raise to prevent neck injury during impact. Toyota vehicles are also equipped with driver and front passenger seatbelt warning sensors which sound an alarm and flash a light on the dashboard if the car is in motion and either the driver or front passenger (if there is one) or both are not buckled in. The faster you drive with the seatbelts disengaged, the faster the alarm beeps. Every seating position also includes a three-point seatbelt system with adjustable driver and passenger seatbelt anchors. The rear seats of all Toyota vehicles are equipped with lower anchors and rear tethers for children (LATCH), which are anchors for outboard rear-seating position car seats. The rear doors are also equipped with child locks to prevent children from opening the door while under way. Every seatbelt is also equipped with pretensioners with force limiters to prevent injury from seatbelts during a collision. There are also automatic and emergency locking retractors on all seatbelts. Every window in Toyota vehicles is pressure sensing. If your child's arm or finger gets stuck in the window while closing it, the window automatically goes down.

As one can tell, Toyota vehicles are crammed full of state-of-the-art safety technology and at no additional cost to you. They are extremely safe vehicles, which is why they have won the NHTSI Insurance top safety award many times over. Toyota is very concerned with the safety of its customers and is the only company to provide all these safety features standard on all models at no additional cost to you. Toyota vehicles are literally the safest vehicles on the road because no other vehicle manufacturer offers all of the security features Toyota does and even if they did, they still would not be up to par with Toyota. No one will ever be able to beat Toyota at anything! My motto is "Toyota's the best and then there's the rest, and that's all there is to it! So get used to it!" There's a reason they are the world's leading automotive manufacturer. If you are looking for a vehicle that's extremely safe, reliable, high quality, and long lasting, then you definitely should go with a Toyota. No other vehicle is even close to beating Toyota quality and safety! Your vehicle is your front line of defense in a collision, so why not choose a Toyota which includes every safety feature on every vehicle, for no additional cost to you? Your family is irreplaceable, so choose a Toyota to protect them.

After reading all of these safety features which Toyota vehicles have, how is it possible for a car to "run away" or accelerate uncontrollably? The answer to that is plain and simple; they didn't. Did you ever notice how the so-called runaways started and stopped at the same time nationwide, even worldwide? The news media channels are biased in favor of "pure" American cars such as the Big 3 (GM, Ford, and Chrysler). All throughout the hype of the so-called Toyota recalls, Ford was also having a ton of recalls. Did the news ever once even mention those recalls? No! Jeep had recalls; did the media announce that? No! General Motors had recalls; did the media ever report on that? No! Just recently, Honda and Nissan both had recalls. Honda's was marginally large, under a million, and Nissan's was over 2 million vehicles. Did the news spend much time talking about those manufacturers' recalls? No! The news talked solely about the Toyota recalls for months and months. The news knew that Toyota is and always will be the world's number one vehicle-manufacturing company and wanted to scare people away from Toyota and make them buy "American" vehicles. Well, that sure did not work!

Every time I log on to Yahoo News or Google News, there are always news reports on how Toyota ranked as the best vehicle manufacturer and how its vehicles outsell any of the competitors. Toyota is not going away and is not going to slow down to give the Big 3 a chance to recover from their costly mistakes and catch up. Toyota has many vehicles which own their category. Toyota builds the Toyota Prius, which owns the hybrid market. When people hear the word "hybrid," they automatically think of a Toyota Prius. When I think of a truck, I think of a Toyota Tundra or a Toyota Tacoma. When I hear the word "van," I envision the Toyota Sienna. Toyota is the best at everything all the way around. No one else can figure out how to build a true hybrid like the Prius, which gets 50 mpg. No one can build a half-ton pickup as good or as capable as the Toyota Tundra. When you compare an "American" van to the Toyota Sienna, there's a HUGE difference in quality and luxury. I work with my dad, who owns Sonic Lube, and I see a ton of Toyota vehicles come in that are well over ten to eleven years old with well over 200,000 miles on them. There's a 1992 Toyota Camry that comes in regularly to have an oil change that has 987,000 miles on it, and it keeps coming back to our shop, closer and closer to a million miles. It has the original motor and transmission. The person who owns the vehicle now is its second owner. She purchased it when it had something like 900,000 miles on it. She said that she knew she was buying a Toyota and had no problem with the mileage on it. Miles on Toyotas just do not matter. Toyota vehicles go on forever and ever and will never let the owner down.

The media were wrong to report on Toyota like they did, and there were several newscasts which proved that the incidents which did occur were all driver error and/or an attempt to get money from Toyota. It is impossible for a Toyota vehicle, such as the Prius, to go out of control. I personally have tried the safety features on my Prius, and they all brought the vehicle to a nice, safe, smooth, and complete stop. There are YouTube videos with other Prius drivers demonstrating this as well. There are millions and millions, if not, billions of Toyota vehicles on the roadways all over the world. People know that Toyota quality, safety, and reliability cannot be beat. *Consumer Reports* still rates Toyota as the best vehicle manufacturer. Toyota's the best, and then there's the rest. That's all there is to it.

Tony Minyon: National Technical Support Manager, Toyota Motor Sales

My story begins in the 1960s, when I was just a little guy growing up in southern California. My family was your usual one with my parents each having a car of their own. My father preferred something large and made in Detroit, while my mom always got the "family mover" that started out with an import little wagon (think of something that is no longer sold in the U.S.).

My dad had the usual problems with his cars that made changing them every four years mandatory, and my mom's had a worse fate. That poor little wagon lasted just a couple of years before it totally began falling apart. The final straw was a hot summer day when it totally died while I was in what I liked to call the "way back" (this is the cargo area that kids liked to sit in back then—without seatbelts), and we had to rely on friends to bring us back home. Our introduction to Toyota started right after that day when we traded in the wagon for next to nothing for a new Toyota.

Our next family movers went from a Corolla wagon, a couple of Celicas, to one of the first Celica Supras (not the speedster that it later became). We were never without at least one Toyota in our driveway. My older sister also joined the fold with a used Celica five-speed when she got old enough to get her license. When my time came to drive, that trusty Celica was the one that I learned to drive a stick with while we drove around the local mall parking lot. All that I could say about that car is that the clutch and transmission must have been supernatural to survive what I had to dish out grinding away at the gears!

Fast forward to 1989, when I was fresh out of college and looking for a job. At the local campus, I was searching the jobs book and sending my résumé anywhere and everywhere that held a remote interest. Toyota Headquarters in Torrance, California, was one of the responses that I got, and I still remember that interview to this day. The job was for a distribution auditor that got to travel all around the country verifying incentive earnings. I kept thinking that this was a great way to see this country while at the same time getting some valuable experience to build upon when I moved on to bigger and better things. The one point in the interview process that always keeps coming back to me was during one of the last rotations: a national manager gave me a valuable piece of advice—he told me that the auto industry was intense and that it will consume you over the years. Wow, was he right.

I have been at Toyota Motor Sales now for over twenty-one years, and I am proud of the company and of the products that we sell (we have been the sales and distribution arm for the U.S. since 1957). I have had incredible opportunities to learn, grow, and contribute as my years progressed. This includes leading the service parts logistics group (transportation) for over five years during a chaotic period with natural disasters, industry growth bottlenecks, carrier consolidations, service parts purchasing from the West Coast and Asia for a couple of years, and investigating advanced technology in audio for a year.

This all leads up to the winter of 2009–2010, when I was called upon to take everything that I had learned and skills that had been taught to help out with the upcoming congressional hearings.

I was in the corporate accessory department (CAD) working on a new audio program when I got a call from our group vice president to come to her office immediately. When I showed up, she told me that I would be on a new project for the next several months in our legal department helping out with the upcoming hearings. She could not add any more except that I had to run over to another building to be briefed by our senior vice president on the project. With the overall outline of the project laid out in front of me, I was asked if I was interested. That took all of two seconds to respond: "Of course, that is what we do."

For the next two months, I was in our legal department (where I never worked before) setting up a project management office to manage the countless requests coming our way. Keep in mind that I am not a lawyer—just a logistics guy that was used to working with trucks, trains, and airplanes during times of stress (strikes, floods, bankruptcies, etc.). What totally amazed me was the amount of cooperation within all of the divisions to pitch in and help. No request was too large and no timeline too short at that time. All I had to do was ask, and the information was supplied just in time

Around February of 2010, things were at a manageable pace again, with the project management on "autopilot," when I got another visit from our group vice president to see how things were going. I spent an hour showing her around the project when the question came up: "What next?" That was my first clue that something bigger was in the pipeline.

During the congressional hearings, Mr. Toyoda mentioned that Toyota would be forming SWAT teams to rapidly respond to any and all quality questions from our customers and vowed to have it staffed and operational by the early spring. This was my opportunity to jump in again and do what I do to set up the program

The following week, I moved again over to the product quality division and was given the task to start the SMART teams (name changed from SWAT). I have the best staff from areas that could contribute in various ways (customer relations, dealer operations, quality, etc.), access to hundreds of field engineers from across the country, communications with the manufacturing companies, and a direct line to senior management without any delay. My marching orders were quite simple: "Look at everything."

The teams were launched in April, and to date we have done thousands of dealership inspections using our standardized protocol, many of which were cross-functional with representatives from sales and quality, engineering, manufacturing, law enforcement, and government with absolutely no surprises. We uncovered most of the stories heard in the news, but we have found no problems with the electronic throttle control as was the initial allegation.

As the team moved forward, we did not stop with investigating unintended acceleration, but have now expanded into other quality claims in order to respond quicker to customers' concerns and to support the quality that we all truly believe is built into every car produced under the Toyota, Lexus, and Scion brands. Mr. Toyoda made the statement to the world during the congressional hearings, and now everyone in the company is backing him up!

The SMART teams will live on long after we have moved on to new positions using the tools that were created during that very trying period, and Toyota will once again be the benchmark of quality in the auto industry.

Chapter 15

My Career and Thoughts

A good head and good heart are always a formidable combination.

—Nelson Mandela

My career began in 1995, but my Toyota story begins in 1992. I had taken a field trip to TMMK while in high school. I know it sounds clichéd, but I decided on that day that I would work for Toyota. I was amazed at the cleanliness of the plant. Naturally, you think that a factory would be dirty and confusing. It is like a city of 7,000 all under one roof. We have our own roads. Ez-Gos and forklifts are going every direction. Just the organization of the place is amazing.

I started working at TMMK on July 15, 1995. It was the week before my twenty-first birthday. Our first week was spent in orientation. During that time, we had the opportunity to learn all about our benefits and what the expectations were. It was our job to be problem solvers. The idea of this excited me. Those that know me personally know that I am never resting. Even if I am sitting, my mind is always thinking, always trying to figure out a better way. My first day in the assembly shop was a meeting with my manager and assistant managers. Mike Hoseus was our manager. He made it a point to make us all feel welcome. He would later leave TMMK to work for a charitable organization. He now has his own consultant company and has also cowritten a book with Professor Liker titled *Toyota Culture*. Our assistant managers were Randy

Brantley and Charles Luttrell. My first team leader was Loretta Galloway. She always bought all of her team members a birthday cake. Here I was, my first week at Toyota, and my team leader was buying a cake with my name on it. I will always appreciate Loretta doing that, because it made me feel welcome right from the beginning.

My first job was to install the left-hand visor and some air ducts. I lost forty pounds in that first month. (I have since gained back sixty). Needless to say, repetitive work is very hard. The only way to describe it is similar to being in an aerobics class for eight hours a day.

One day, I was really struggling on the left-hand visor process. Randy walked by and noticed that I was working hard and struggling to keep up with the pace of the line. He stopped and helped straighten up my flow rack. I was amazed that he did this and appreciated it very much. It actually gave me a chance to catch my breath. That was my very first lesson from an assistant manager on Toyota's value of respect for people. My group leader at the time was named John. He is a good man. I learned a lot from him. While I was new, he had me visit with him one day. I thought I had done something wrong. He sat me down and asked me what my career goals were. At the time, I did not think much of it. It would be years later before I would realize that it wasn't the idea of him worrying about my future but the fact that he was taking the time to get to know me. He is a great leader. John, I have carried the lessons with me that you taught fifteen years ago.

After leaving the trim section, I would go to dayshift and work in the chassis section. I had many good leaders while in this group. It would be a few years before I would get another great leader to learn from, though. Kenny Clemmons was one of the best leaders I ever had the chance to work for. He truly cared for his employees on a personal level. Kenny was my group leader when I got promoted. Tragically, Kenny was taken from us in a terrible car wreck. He left a strong legacy within our walls. He would have over 600 people attend his visitation and funeral. The support given to his family was a testimony to the kind of life he led and his character as a human being.

I spent eight years in the chassis section, the last three as a team leader in MA420. My confidence as a leader was raised while in this group. I made it a point to take all I learned and apply that to my group and team members. I would stock all their parts. I would clean the area. It was the team members' jobs to build the cars. It is a team leader's job to help them be successful. After all, the cars being built to high quality are what pay all our paychecks. My group leader at the time was Tony Hendrichs. Tony graciously shared his story earlier in the book. I loved working for Tony on the production floor. Tony has the keen ability to know his team members' strengths and then use those to the benefit of the group and company.

After leaving chassis, I went to the final section. I wanted something different and wanted a new challenge. This ended up being the biggest challenge I had in my career thus far. I thought I was a pretty good team leader. My time in chassis had not really prepared me for what I was about to be a part of. Quality is important in chassis. After all, if the work isn't done right in chassis, it could cost someone their life. The parts were big, and room for error was zero. We had many systems and tools that would prevent most defects, but the team member's judgment always has and always will play a big role in our success.

Final section, on the other hand, dealt with a lot of plastic parts and the seats. All the work was heavy in chassis. Everything in the final section had to be done with finesse, or you would cause a stress mark or scratch. The seats are not as easy to install as one would imagine. I have ran close to 100 processes in my career, and I can tell you without a doubt that the rear seat install process is one of the toughest I have ever done. They have to be placed just a certain way, and the bolts holding it together have to be a certain torque. This is called a "Delta C process." Labeling something a Delta C is a way for us to share the importance of the work. If it is labeled a Delta C, then that part could cause someone their life if not done correctly. Needless to say, the attention scale is very high on all these processes. The last thing any of us wants is to know that our lack of attention cost someone their life. We all realize that responsibility and take it very seriously. My time in final lasted for one year before I had the opportunity to move to the safety team. My time in final taught me about the kind of person I really wanted to be. I would take care of my team members and friends no matter what it cost me personally or professionally.

Now, I thought I would share some inspiration of what has gotten me to this point in my life. These people are not Toyota employees but they have made an impact to the company through me working here, and I hope sharing this insight will have an impact on you as well. I believe in the idea that we say, "Thank you," and pay tribute to those who came before us. It is not every day that a factory worker from Kentucky decides to write a book, so I am sharing stories of my family and friends and what I have learned from them.

"Rocking Chairs, Patchwork Quilts, Hard Work, Values, and Friendship"

The Turners

My parents moved from eastern Kentucky to northern Kentucky before I was born. My dad worked for the same company for thirty-four years retiring last year. He is one of the greatest people I know. He is a quiet, humble man who loves his family. Even though my parents lived near Cincinnati, Ohio, they would instill the values that they grew up with.

To understand those values, you first must understand how times were in eastern Kentucky during the 1950s and 1960s. As my mom shared, electric was not installed in the one-room schoolhouse until the Johnson administration. They remember when plumbing was first installed in their houses. Before that, everyone had an outhouse. Imagine in the middle of winter having to put on your coat and shoes and walk outside in the snow just to use the restroom. We take many things for granted now. Times were truly hard.

Jobs were unique and different, or you had to travel to find good work. My Papa Greenberry Turner was pretty much an orphan at six months. He was the son of a seventy-three-year-old Civil War veteran. His father died soon after he was born. His mother remarried but kept the original house. Papa would be raised by his brothers and sisters. He started working in a coal mine at the age of fourteen. It was his job to work the water pump and also to pump the air into the mine to provide oxygen to the men working. Can you imagine putting that much responsibility on a fourteen year old today? This job taught him the importance of doing the job right.

He married in the mid-1920s to my Mama Bertha Turner. She would also have a hard life. She loved her family and was a devoted wife and mother. This devotion would be passed along to her family for generations to follow. She would cook the food for the children at the schoolhouse. She would make the meal that Lady Bird Johnson would eat while making her visit. They instilled in their children many values such as responsibility, hard work, appreciation for what they had, and devotion, which were evident and live on in their kids and grandkids today. Papa would eventually begin a unique job. He would deliver mail on horseback. He would do this job until the mid-1960s and would retire. He would begin woodworking, making rocking chairs. It is unknown how many of these chairs he made, but I would venture to say that it is in the 10,000 range, and they are in every state.

I was in a gift shop the year after he passed away and saw one of his chairs for sale. I knew it was his because of the workmanship and the standardized way the legs were turned on the lathe. When I first saw it, I decided not to buy it. I thought, "I already have one." Now that I have kids, not buying that chair on that day is one of my biggest regrets.

The Bowlings

As you could tell, my mother loved her parents. They were both outstanding people. They loved their kids. My Papa Walter Bowling would travel to Cincinnati to find work. He would drive up there every week and take other men with him. The men would pay him gas money. This would be a way he could help others and provide added income to raise his large family. This placed many responsibilities on the siblings. My two uncles were the oldest of the kids and would take on many responsibilities that most teenagers would not dream of doing today. My mom was taught teamwork and devotion to family throughout her life. Teamwork was needed for survival. I remember, growing up, when we would go back to Jackson I would go with Papa to help others. One story that I remember was assisting him in helping a neighbor move a hog. Now that I have gotten older, I realize that it was much more than the nasty job of moving the pig. It was about helping a neighbor do something that no one enjoyed doing. The very danger of the job caused others to come together to get the job done. Supporting and helping our neighbors. This is a value that we all need to have.

Both my grandmothers loved to make quilts. This would be a time to get together with other ladies in the community. I am sure they would set around the quilt rack and talk and stitch. This would also teach teamwork. The ladies found that if they worked together, they could get a quilt done much quicker. It would also improve communication in the community at the same time. My Granny Sarah Belle Bowling would work together with many ladies on these quilts. They made one and gave it to Lady Bird Johnson on her visit to the schoolhouse. My parents have a picture of each of their mothers with Lady Bird on this historic visit. I would guess that the first lady took more away from her visit than she gave.

When I began this book I wanted to focus on the Toyota Way principles and how those promote loyalty. I wanted to explain why Toyota chose Kentucky. The reason they chose our state is because of the circumstances that created our values.

For any company or person to be successful, we have to have the values of my humble grandparents: teamwork, communication, responsibility, hard work, appreciation, devotion, and helping a neighbor, even if the job isn't easy.

The Teacher

There is a saying told from most Japanese trainers that if the student hasn't learned, then the teacher hasn't taught. I know this to be a true statement in both work life and personal life. Eugene C. Keith was my wood shop teacher at Simon Kenton High School. He did something that was very special. He would search out and find the students who were overlooked. We didn't play sports. In most students' cases, we were shy and would keep to ourselves.

I remember the first day I met the man, he decided that I was one of his chosen students. He told me within the first week that he was going to break me of my shyness. He encouraged me to get involved with a club at school called the Technology Student Association.

This club would be an opportunity for me to get involved and excited about going to high school. I would become an officer of the club every year. On my senior year, I would be the president. Leading the way of 100 of my fellow students to become state champs and go on to the national competition. In my sophomore year, Mr. Keith would take three of his students to a leadership training camp. I was lucky enough to be one of those students.

My senior year, he encouraged me to run for a state officer. I had to get in front of 200 students from across the state and give a speech. The idea of doing this terrified me. He made me get up there anyway. He knew this was a step in a journey that he began four years before. He knew if I could do what he was asking, that I would break my shyness. I didn't win, but I now have good communication skills.

Mr. Keith always strived for his students to succeed in life. He took us on a field trip in 1992 to TMMK. I didn't tell him at that time, but I had decided on that day that I would apply for the job. I remember telling him my senior year that I wasn't going to college. I was going to work at Toyota. He had a confused look on his face. He had assumed that I would go to school and take over for him when he retired. He hadn't shared this plan with me. If I would have known, I would probably be teaching now. I could not or would not ever tell him no. He was loyal to his students, and respect is what he earned.

Mr. Keith passed away on February 13, 2003. He had been retired for around five years. He would have complications from a staph infection. It was hard to see this humble, great man spend his last months in the hospital. I would go there nearly every week just to sit and talk with him. He asked me once if he had really made a difference. I guess when you know your time is up, you really begin to reflect upon your life. I remember like it was yesterday. I told him that he had affected every life of every student he taught. We were the students who were overlooked, and he always expected the best out of us. He always gave us his best. This man changed my life, and for that I will forever be grateful.

For a company or leader to truly be successful in life, we have to follow the leadership of Mr. Eugene C. Keith. He realized that he was successful when he searched out and challenged the students who were overlooked and then was devoted to the success of those students. There are thousands of people who had him as a teacher. I am confident that the older they get, the more they realize exactly what he was doing all along.

Chief Quality Officer and Former TMMK President Steve St. Angelo

A few years ago, I met Steve because one of my team members needed help and I could not do it alone. He needed someone with influence to help him. I had exhausted other possibilities, and finally I called Steve. I talked to a great lady named Rebecca Lucas who said she would pass along the message. I really didn't expect to hear back from him. My expectations changed the very next day as I was on my way to work and my wife called my cell phone to tell me that the president of Toyota had just called the house asking for me.

I called him back and explained the situation. He helped my friend, and in doing so he created a very loyal employee. Six months later, I was invited to a luncheon held by him. This lunch was not advertised or promoted in any way. He wanted to recognize some of the team members that had done good deeds for each other. The people in the room went above and beyond for a coworker. Personally I think we were all doing what anyone would do, but he wanted to pay a tribute in some way. One thing that makes him such an effective leader is his recognition of his team members' good works.

Steve started off on the factory floor. His dad worked as a maintenance supervisor for General Motors. Steve would get his start delivering food on a cart to the assembly workers at GM. He has told me that he learned great interpersonal skills while doing this job. He would learn to listen. He would see for himself what issues the workers had to deal with. He would begin college and eventually get his master's degree in manufacturing management from Kettering University. He made the most of his opportunities, worked hard, and always reached for the stars.

He believes that good morale and development of his workers will produce great results. So often, this is the case in the world. If a person is positive and speaks on the "good things," then most people will respond likewise. Years ago, I would have never imagined that I would be putting a book together to share the Toyota story. My devotion to the company is based on him helping my friend.

Every person in this world has influence in someone else's life. We all have to make a conscience decision in how we want that influence to be used. Thankfully, our chief quality officer uses his influence for the common good. The struggle with being a leader is that everyone doesn't see the good. Some people only see the bad. I can assure you that when the going gets tough, it takes people like Steve to ensure the best thing possible for everyone is what happens.

Just like Mr. Cho, Steve believes in the idea of being on the floor. I am sure that he would be on the floor all the time if he were able to. The team members appreciate him taking the time to stop and talk. He has the ability to talk to community leaders or factory workers and make us all feel like we are an important piece to the puzzle. Every year, he goes to the day care and celebrates the holidays before winter shutdown.

One thing I have learned is that if a company wants to be successful, they have to hire people with the same values of the company. Toyota has done that with the hiring of this man. I have also learned that any successful person must have a supportive spouse. I have had the pleasure of meeting Michelle St. Angelo. She is a great person, a supportive wife to Steve, a wonderful mother to their two daughters, and a great friend to everyone who has the privilege to know her. They have traveled the world together in Steve's thirty-six years in the auto industry, and I think it is safe to say that Steve would admit that his success would not have been possible without the love and support of his wife and family. They are great role models for anyone who wants to do more in life and still be able to keep family as the priority.

Steve's job now is to be the voice of the customer. He is responsible for all the quality in Toyota products in North America. He has a direct line to Akio Toyoda and can call him anytime with an issue. Toyota will place the customer first because that responsibility falls under Steve St. Angelo, and I can't think of a better person to have that responsibility.

What Is the "Magic" of TMMK?

That is the $1,000,000 question that every manufacturer asks in North America. Many companies come here. As Mr. Goetz stated, they see our equipment and our kanban system but they can't grasp the culture. What makes it unique? Many consultant companies have spawned from former TMMK employees, and most likely another will begin when I retire from here. It has been said that the foreign auto makers take advantage of the people and cheap wages from the South. While it could be seen this way, I as a Kentuckian want to discuss this further.

In the mid-1960s, President Johnson declared a War on Poverty. President Kennedy had done so before he was tragically assassinated in 1963. President Johnson focused on the people of the Appalachian mountain areas. They would do many good things like provide utilities to an area of the country that had long been forgotten. They couldn't have realized in the 1960s that it would take the presence of a foreign auto maker to help their goals become a reality.

Toyota did something amazing for our region and our people, just like Ford has done for the Louisville, Kentucky, area and General Motors has done for the area in Bowling Green, Kentucky. They give us purpose. They give us security. Most importantly, they give us pride. We are proud of what we do. Just like our ancestors were proud miners, loggers, and farmers.

Kentucky has always been a state that provided the country with much-needed resources. In the 1800s, it was timber. In the early 1900s and today, it is coal. We also provide our country with the bestselling car in the United States. It is built by Americans for Americans. Our company directly provides over 30,000 Americans with manufacturing work with 7,000 direct jobs in Kentucky alone. There are 100 Tier 1 suppliers located in our state that provide an additional 18,000 jobs. To me, that sounds like we are making quite a contribution to our great country.

This book has not been negative in any way. We have not said a bad thing about any other manufacturers. That was never the intention. The idea of this book stemmed from the hope that we as Toyota employees could say, "Thank you," to the corporation that came here twenty-three years ago and gave us all the opportunity of a lifetime. We wanted to share that good things happen here and that we are just normal hardworking human beings just like the people reading this book. While we are not perfect in any way, Toyota would be the first to say that. Part of our "magic" is the idea that we know we aren't perfect. We always have room to improve and are never resting. Once we think we have begun to reach perfection, we start over again.

Part of our "magic" comes from quality circles. It is my guess only that a lot of consultant companies out there do not teach about quality circles. The teamwork and continuous improvements that are caused by them are "magic."

Our outreach program creates the magic. How many large companies do you know that do this sort of thing? I have thought of ways to describe it. I call it small company values being used by a corporation. In my opinion, it is a key ingredient to the culture we have at TMMK. Thankfully, I have been blessed and not had the need for this fund. I have many friends who have needed it, and it is always there to help.

Diversity creates the magic. While we have always been a diverse work culture, we do not only look at the color of skin or nationality. Our diversity stems from our personal experiences and beliefs. I have grown a lot as a human being while working here just for the opportunity of getting to know so many people from so many backgrounds.

Respect for people creates the magic. Some would say that this is a term thrown around too loosely. I believe that Toyota at its core has respect for its team members. Respect isn't just about being nice or giving people whatever they want. It is about doing what you say you will do and allowing people to grow within the company. Mr. Ohno is a great example of this. He was born in China and started working for a Japanese company on the production floor and was given the opportunity to better the company. In doing so, he changed the world. Every company is trying to go to a lean-manufacturing model. Mr. Ohno was given the opportunity from the Toyoda family and took the challenge and ran with it. Now that is magic.

Our stance on the environment creates the magic. Eight of the ten Toyota facilities in North America have gone to zero landfill. We segregate our trash and recycle as much as possible. The money made from the recyclables goes toward our benevolent fund. We are using a positive action to promote a positive benefit to the team members.

Working for a company that is forward thinking about the research and development of the hybrid car creates the magic. It is a good feeling to know that the success of the Prius and now our Camry Hybrid has helped promote the idea that hybrid cars can be stylish and roomy and still help the environment. The announcement that Toyota was partnering with Tesla Motors proves that Toyota is looking for other opportunities to improve the world through hybrid cars.

Toyota's belief in helping our communities creates the magic. There are countless ways that our company helps our communities. Just this past year, my Cub Scout troop worked with nearly 100 Toyota employees on National Public Lands Day. We worked at Waveland Historic site in Lexington, Kentucky.

The company's investment and reinvestment back into the U.S. economy create the magic. Toyota has ten factories within the United States. When other companies are leaving the United States for cheaper labor in other countries, Toyota is truly showing that it is a global company by building and designing the cars where they are sold.

Our leadership creates the magic. From our first leaders, Kanayoshi Kusonoki and Mr. Fujio Cho, we knew that we would be given every opportunity. Mr. Cho was trained by Mr. Ohno. Toyota wanted us to succeed so much that they sent two of their greatest leaders to Kentucky. Mr. Cho was given a huge challenge and succeeded. That success is still felt today. Now we have our current president. Wil James is leading the way for TMMK to help us reach our goal of winning the JD Powers quality survey again. He and chief quality officer and former TMMK president Steve St. Angelo have one of the greatest challenges of any leader at TMMK thus far. With the economy the way it is, he has to make tough decisions. No one likes to have their pay cut or miss their bonus. Our leadership in Japan and Mr. St. Angelo are making the decisions that no one else wants to make. Every decision made now affects our community. Imagine going out to a restaurant to eat. You look around and see all these families. You know that the decisions you make affect the lives of at least 30% of the families eating around you. If you had to lay off someone, how could you choose? Is one family less important than the next? Of course not; that is why our leadership has chosen to keep all its employees working. Having the courage to make those decisions and making the right one create the magic.

While we like the pay and benefits, it isn't only about those things. It is about loyalty, a mutual loyalty between employee and employer. It is about looking out for your fellow team members.

In closing ...

Gandhi was quoted as saying, "My life is my message." Every day I say two prayers: they are my guiding principles and goals each day. At night I pray the prayer of Jabez. Jabez is only mentioned in the Bible in two verses. He asked that the Lord bless him and his house and extend his territory. He was asking that God extend his territory so that he could use it in his works for the Lord. My second prayer is to ask God to give me a "perfect day." Coach John Wooden once said, "You can't have a perfect day without doing something for someone that they can never repay you." These are my guiding principles and my goal each day.

Earlier in my chapter, I talked about my grandparents and family. I talked about what I learned from them. I see the Toyota Motor Corporation in the same light. We are a descendant of the corporation in Japan. We are an American company. We are a good corporate citizen. We support our military and pay our taxes. The company gives to charitable causes, especially education. The company provides many Americans with work. We are proud of our Japanese heritage, though. With that heritage comes the knowledge of building a good-quality car and working for a company that bases its decisions on following our values and principles.

Doing this book has been very rewarding. I excitedly told Mr. Raymond Bryant that I have learned something new with each addition that I have gotten. This is the team members' opportunity to remind everyone of our contribution to society and each other. Each person directly or indirectly has an affect on other human beings. If this book and Toyota teachings can help companies to succeed and Americans to keep working, then all the work has been worth it.

I have had the opportunity to get back in touch with old friends. I had asked the team members to share stories of their motivation, determination, understanding, and kaizen. I had asked our leadership to share stories of their knowledge. This book has been a journey that I have been proud to be a part of. It is a perfect metaphor for what our company is all about.

Teamwork: A group of people working together to achieve a common goal. Everyone in this book was asked by me personally to participate. They all had a story to share.

I am a student and teacher. Even though I do not have any degrees, I take the opportunity to learn all I can from everyone I come in contact with. There is always a new lesson to be learned and a new opportunity to share what I have learned with others. This is the Toyota Way. We learn from our mistakes, and when asked we help others to not make those same mistakes. It makes us all more efficient.

I hope this book has given you a sense of our pride in our work and our commitment to our coworkers, our company, and our customers. I hope it has given you an understanding of the Toyota production system and reassurance that you are getting a great car built by good people.

If Toyota has a magic formula, it is that everyone in the company has the same goal. From the president to the team members, we are "One Team on All Levels."

Index